DATE DUE

PRINTED IN U.S.A.

Information and Innovation

Medical Library Association Books

The Medical Library Association (MLA) features books that showcase the expertise of health sciences librarians for other librarians and professionals.

MLA Books are excellent resources for librarians in hospitals, medical research practice, and other settings. These volumes will provide health care professionals and patients with accurate information that can improve outcomes and save lives.

Each book in the series has been overseen editorially since conception by the Medical Library Association Books Panel, composed of MLA members with expertise spanning the breadth of health sciences librarianship.

Medical Library Association Books Panel

Lauren M. Young, AHIP, chair
Kristen L. Young, AHIP, chair designate
Michel C. Atlas
Dorothy C. Ogdon, AHIP
Karen McElfresh, AHIP
Megan Curran Rosenbloom
Tracy Shields, AHIP
JoLinda L. Thompson, AHIP
Heidi Heilemann, AHIP, board liaison

About the Medical Library Association

Founded in 1898, MLA is a 501(c)(3) nonprofit, educational organization of 3,500 individual and institutional members in the health sciences information field that provides lifelong educational opportunities, supports a knowledgebase of health information research, and works with a global network of partners to promote the importance of quality information for improved health to the health care community and the public.

Books in the Series

The Medical Library Association Guide to Providing Consumer and Patient Health Information edited by Michele Spatz
Health Sciences Librarianship edited by M. Sandra Wood
Curriculum-Based Library Instruction: From Cultivating Faculty Relationships to Assessment edited by Amy Blevins and Megan Inman
Mobile Technologies for Every Library by Ann Whitney Gleason

Marketing for Special and Academic Libraries: A Planning and Best Practices Sourcebook by Patricia Higginbottom and Valerie Gordon

Translating Expertise: The Librarian's Role in Translational Research edited by Marisa L. Conte

Expert Searching in the Google Age by Terry Ann Jankowski

Digital Rights Management: The Librarian's Guide edited by Catherine A. Lemmer and Carla P. Wale

The Medical Library Association Guide to Data Management for Librarians edited by Lisa Federer

Developing Librarian Competencies for the Digital Age edited by Jeffrey Coghill and Roger Russell

New Methods of Teaching and Learning in Libraries by Ann Whitney Gleason

Becoming a Powerhouse Librarian: How to Get Things Done Right the First Time by Jamie Gray

Assembling the Pieces of a Systematic Review: A Guide for Librarians edited by Margaret J. Foster and Sarah T. Jewell

Information and Innovation: A Natural Combination for Health Sciences Libraries edited by Jean P. Shipman and Barbara A. Ulmer

Information and Innovation

A Natural Combination for Health Sciences Libraries

Edited by
Jean P. Shipman
Barbara A. Ulmer

ROWMAN & LITTLEFIELD
Lanham • Boulder • New York • London

Published by Rowman & Littlefield
A wholly owned subsidiary of The Rowman & Littlefield Publishing Group, Inc.
4501 Forbes Boulevard, Suite 200, Lanham, Maryland 20706
www.rowman.com

Unit A, Whitacre Mews, 26-34 Stannary Street, London SE11 4AB

Copyright © 2017 by Medical Library Association

All rights reserved. No part of this book may be reproduced in any form or by any electronic or mechanical means, including information storage and retrieval systems, without written permission from the publisher, except by a reviewer who may quote passages in a review.

British Library Cataloguing in Publication Information Available

Library of Congress Cataloging-in-Publication Data

Names: Shipman, Jean P., editor. | Ulmer, Barbara A., 1957- editor. | Medical Library Association.
Title: Information and innovation : a natural combination for health sciences libraries / edited by Jean P. Shipman and Barbara A. Ulmer.
Description: Lanham : Rowman & Littlefield, [2017] | Series: Medical Library Association books | Includes bibliographical references and index.
Identifiers: LCCN 2017004377 (print) | LCCN 2017028699 (ebook) | ISBN 9781442271425 (electronic) | ISBN 9781442271401 (hardback : alk. paper)
Subjects: LCSH: Medical libraries—Aims and objectives. | Medical libraries—Technological innovations. | Medical librarians—Effect of technological innovations on. | Libraries—Space utilization. | Medical libraries—United States—Case studies. | Medical innovations. | MESH: Diffusion of Innovation | Libraries, Medical | Information Dissemination | United States
Classification: LCC Z675.M4 (ebook) | LCC Z675.M4 I435 2017 (print) | NLM Z 675.M4 | DDC 026/.61—dc23
LC record available at https://lccn.loc.gov/2017004377

∞ ™ The paper used in this publication meets the minimum requirements of American National Standard for Information Sciences Permanence of Paper for Printed Library Materials, ANSI/NISO Z39.48-1992.

Printed in the United States of America

Contents

Figures		ix
Tables		xi
Foreword		xiii
Christopher Wasden		
Preface		xvii
Jean P. Shipman and Barbara A. Ulmer		
1	Deep History: Creativity, Innovation, and Libraries *Joseph Lucia*	1
2	Innovation Cycle and Information Applications *Jean P. Shipman, Tallie Casucci, and Spencer W. Walker*	13
3	Synapse: A Place Where Ideas Collide and Collaborations Congeal *Jean P. Shipman*	23
4	Making the Makerspace: The Nexus Collaborative Learning Lab *Jennifer Herron and Kellie Kaneshiro*	33
5	Supporting Institutional Strategic Directions and User Needs through Library Collaborative Spaces *Mary Joan Tooey*	43
6	Gary L. Crocker Innovation and Design Laboratory, University of Utah *Jean P. Shipman and Timothy Pickett*	55
7	Library Makerspace Programs: Bringing Together Space, Services, and Staffing *Elliot Felix and David Woodbury*	67

8	Interview with Roger Altizer, PhD, and José Zagal, PhD (March 2, 2016) *Jean P. Shipman and Barbara A. Ulmer*	77
9	Information Needs of Medical Digital Therapeutics Personnel *Tallie Casucci*	89
10	Medical Innovation Competition Information Support *Erin Wimmer, Tallie Casucci, Jacob Reed, Nathaniel Rhodes, Benjamin Fogg, Thomas J. Ferrill, David Morrison, Alfred Mowdood, Darell Schmick, Mohammad Mirfakhrai, and Peter Jones*	99
11	Innovation Space Drives Need for Librarian Expertise *Jean P. Shipman and Tallie Casucci*	117
12	Applying Innovation to Patient Education and Behavior *Roger Altizer Jr. and José Zagal*	131
13	e-channel: A Platform for Disseminating Innovators' Outputs *Christy Jarvis, Chad L. Johnson, and Jean P. Shipman*	143
14	Building Innovative Products via Successful Partnerships *Nancy Lombardo and Kathleen Digre*	157
15	Educating Innovators: *The Innovation Vault* *Barbara A. Ulmer and Christy Jarvis*	169
16	Information and Innovation: What Does the Future Hold? *Jean P. Shipman and Barbara A. Ulmer*	177

Index	185
About the Editors and Contributors	189

Figures

Fig. 1.1	Stoa of Attalos in Athens.	3
Fig. 1.2	1857 British Museum Reading Room.	6
Fig. 1.3	Graduate Reading Room, Suzallo Library, University of Washington.	7
Fig. 1.4	Reading Area, Main Branch of Seattle Public Library.	8
Fig. 3.1	Final Synapse Design.	26
Fig. 3.2	Synapse Café.	28
Fig. 3.3	Synapse.	31
Fig. 3.4	Synapse Donor Wall and Lounge Area.	31
Fig. 3.5	Synapse Group Bay Area.	32
Fig. 4.1	Nexus Mobile Furniture.	37
Fig. 4.2	Nexus IQ Wall.	39
Fig. 5.1	Presentation Practice Studio.	45
Fig. 5.2	Innovation Space. Health Sciences and Human Services Library, University of Maryland.	49
Fig. 6.1	Crocker Lab.	57
Fig. 6.2	CNC Mill.	59
Fig. 8.1	The GApp Lab.	78
Fig. 10.1	Seven Areas of Innovation.	110

Fig. 13.1	Original e-channel Interface.	147
Fig. 14.1	NOVEL Project Timeline.	158
Fig. 14.2	NOVEL Users Originate from Many Countries.	165

Tables

Table 7.1	Types of Makerspaces.	70
Table 10.1	University of Utah Medical Innovation History.	101
Table 10.2	Bench to Bedside Statistics.	109
Table 15.1	Topical Playlists.	175

Foreword

Christopher Wasden

So often when I meet creative people who want to invent the newest and greatest digital health devices, apps, games, or services, I am surprised at how little research they have executed. They have not looked at the competitive environment and what other apps, devices, and digital health solutions already exist and focus on the problem. They have not done an in-depth analysis to get to the root cause of the problem, which means they do not understand the failures that create the problem, the pain points that stakeholders experience, and the tension it generates. They are so convinced that they have a brilliant idea and solution, despite the lack of rigorous research and analysis, that they launch into developing the solution. Because they did not do their homework, the result is that it nearly always fails. I call this *mindless failure*. Innovation is fraught with failure; therefore, great innovators apply fast, frequent, frugal failure in order to rapidly innovate and thereby limit their investment in failure and maximize their return on failure. Failure is the genesis of innovation, but it needs to be *mindful failure*, not *mindless failure*. We can achieve this only if we do our homework, create hypotheses, conduct research, test hypotheses, learn from the inevitable failures, and then iterate and improve our novel ideas. Nowhere is this more necessary than in the newly emergent and rapidly evolving digital health space occupied by wearables, mobile devices, apps, and game technologies transforming consumer and clinician behaviors to deliver better health outcomes at lower costs.

When I joined the University of Utah (U of U) in 2013 as the new executive director of the Sorenson Center for Discovery and Innovation, I began to explore who was part of the innovation ecosystem at the U of U. Having spent the previous seven years as the global health care innovation leader at PwC, a global professional services firm, I had become focused on digital health innovation and was eager to find those at the U of U who shared these same interests. Two of the key attractions to coming to the U of U was its top-ranked Entertainment Arts and Engineering (EAE) video game development program and its medical school, which is always ranked in the top 10 for quality and outcomes. The business school, which I joined as a member of the Entrepreneurship and Strategy Department and where my innovation center is housed, regularly falls within the top 20 entrepreneurial business schools in the country. I envisioned that if we could create a trifecta of business, gaming, and health, we could do many things here at the U of U that would be tough to achieve elsewhere. To test this hypothesis, I immediately began to lead an effort to enlist like-minded people from across the U of U to create from scratch the Games4Health Challenge, a global digital game design competition.

The logical partners to start this competition were people from the EAE, medical school, and other entrepreneurial programs. But to my pleasant surprise, this also included the director of the Spencer S. Eccles Health Sciences Library, Jean Shipman. This partnership with Jean and her team, including an innovation librarian, has become invaluable as we have combined the need to innovate with the requirements to do it in a mindful way to avoid mindless failures. She had her team support many aspects of our Games4Health Challenge competition, including students and sponsors that need research support, and she has also provided services to archive information, resources, and past competition submissions so that people can learn from this content in a mindful way. In our third year of this competition, we had more than 400 students from 71 universities and 12 countries compete for $60,000 of prize money in five different health and well-being categories—by all measures, this has been an outstanding success.

We have since expanded our collaboration in other innovative areas of not only research but also implementation. The professionals in our health sciences library have not only research talents but also physical space, access to students, and the ability to help incubate ideas and test them out to see how we can more effectively iterate our innovations. For example, Jean and her team have been a driving force in developing our U-Bar, which, like the Apple Genius Bar, is a place where experts (in this case, librarians and students) help patients and clinicians learn about a curated app formulary, download apps, and deliver provider-prescribed apps to address health and wellness problems. Our librarians not only helped to research apps to create an up-to-date curated app formulary but also provided the physical space for the U-Bar and the human resources to plan, train, educate, and deliver the associated services.

In our dynamic and fast-paced world that is driven by disruptive innovations that demand rapid responses and adaptation in creative ways, we do not have the time to continually reinvent the wheel, to fail where others have already failed, or to continually experience maladaptive tensions through mindless failure. We must engage our research and information experts in our libraries to help us understand where, what, how, and why innovative efforts of others have failed in the past so that we can learn from others' failures and apply fast, frequent, frugal failure in our own creative efforts. By partnering with our libraries, we can accelerate our innovative efforts, become digital health innovation leaders, and deliver better solutions that combine disparate talents and capabilities in an interdisciplinary manner to transform the practice of health care and achieve the triple aim. I have been both surprised and gratified at how our health sciences library and librarians continually step up to help lead and drive our innovation efforts forward here at the U of U. I think they provide a great example to others about the role of not only research but also implementation and execution in moving new ideas and plans forward. Everyone who reads this book will have visions of the possible expanded and will need to rethink some of their own mental models of their roles in innovation so that they can expand both their influence and their impact in enabling new value-creating novelty among their stakeholders.

Preface

Jean P. Shipman and Barbara A. Ulmer

Innovation—what a rich and complex word, one that appears almost everywhere in today's world. Innovation has so many meanings and so many applications. There are innovations in processes, innovations in product development, innovations in social entrepreneurship, and innovations in business models. This book focuses on innovation as it affects health sciences libraries and librarians. What is it that health sciences librarians can contribute to the growth of innovation and to those who innovate in health care? How can information impact the success of product development and promotion? There is an untapped reservoir of partnerships just waiting for health sciences libraries and librarians to engage in with innovators; this book illustrates the voluminous number of opportunities awaiting those ready and eager to partake. While there are many books about innovation itself, there are not any that the editors are aware of that directly address innovation and librarians, especially health sciences librarians. This book takes care of that by providing an overview of how innovators benefit from a partnership with librarians when developing their products.

With the decline of clinical income and research dollars, many academic health sciences centers are looking toward innovative product development as their new income source. Health sciences librarians and libraries can partner with these revenue-generating innovators to offer invaluable services, evidence, training, dissemination venues, and attractive collaborative physical spaces equipped with the latest tools, such as 3D printers, body scanners, models, and video monitors. Many health sciences libraries are transforming their physical spaces into collaboration or makerspaces to spark innovation and discoveries. This book highlights health sciences libraries that have done so to enable others to learn more about what professional benefits result from such collisions of information and innovation. Transferring the knowledge of librarians who have progressed down this path to others is the key goal of this book.

The first chapter of the book, written by Temple University librarian Joseph Lucia, provides the historical context of how libraries originated as centers of collaboration and idea sharing. They were often open-air facilities that offered groups a place to congregate to exchange ideas and to ponder complexities to derive relevant solutions. As serving as buildings of stored print information is becoming less necessary in the digital age, libraries today are returning to this primary mission of being a center for concept development.

The next chapter, by Jean Shipman, Tallie Casucci, and Spencer Walker discusses the cycle for the development of innovative technologies in the health care arena, such as medical devices. The authors cover how innovative efforts gain traction with the backing of evidence and key knowledge. They highlight university resources that are available, including experts to help teams of innovators understand the context within which they are developing products, as well as how to identify prior art. Information and innovation are natural partners, as discussed in this overview chapter about the innovation life cycle.

Libraries as collaborative and innovative engaging spaces—that's what the next six chapters cover (chapters 3 to 8). Case studies of three health sciences libraries are offered by three different health sciences librarians. How their libraries came to be involved with innovation is explained, as are the changes enacted that enabled physical library space to be redefined and allocated. Resulting partnerships are also highlighted, as are lessons learned with journeying down this path. One chapter provides an overview of how a library mailroom was transformed into a medical device–prototype fabrication laboratory, complete with staffing to direct innovators and students with safety measures and materials to bring ideas to a physical form. Chapter 7 is written by a consultant who has worked with numerous libraries to reimagine their spaces in order to lend support to innovation in the truest sense of the word. Within many of these case study chapters, photographs help to envision how the spaces have been transformed and how they enable innovators to be highly productive.

In the next two chapters, digital medical therapeutic game creators are interviewed to learn about their information needs and preservation desires for what is created within their workplaces, and an innovation librarian explains how she collaborates to meet these gamers' identified information needs.

At the University of Utah (U of U) and other universities, a variety of educational programs[1] and student team competitions[2] encourage students to learn more about entrepreneurship, innovation, and the business models and planning that accompany the start of a company. Two of the U of U student competitions, Bench to Bedside and Games4Health Challenge, engage librarians from all three U of U libraries (academic, law, and medical) to learn more about prior art, patents, business plans, evidence-based health practice, and other topics as they develop their games and devices. The three libraries formed a team, called the U of U Libraries Innovation Team, to offer students mentors and experts who can guide them in developing the components required by the competitions. Chapter 10 offers an insight into the value that innovation students place on the assistance they obtain from librarians.

As space is shared with innovators, collisions among them and health sciences librarians working directly with them have provided new roles and responsibilities that continue to partner information and innovation as a natural duo. In chapter 11, an innovation librarian and a health sciences library director share how their lives have changed by working in close proximity to innovators, students, faculty, and community industry mentors.

In the next several chapters, emphasis is placed on examples of products that are generated by innovators, gamers, professional societies, and librarians collectively and how information lent a helping hand in formulating the resulting effective and verified products. Gamers share how they work with faculty to collectively design games and apps targeted for patients and consumers. Librarians and staff at the U of U's Spencer S. Eccles Health Sciences Library (EHSL) have been innovative in creating e-channel, a Web-based innovation-themed multimedia portal that captures and documents the productivity of innovators across the country and internationally and illustrates their contributions for university faculty promotion and tenure portfolios. NOVEL, another multimedia product developed as a partnership between a professional society and EHSL, is highlighted, along with its development stages, in chapter 14. The following chapter outlines how a collection of videos with numerous innovators was gathered and described to enable easy discovery of the video content as well as new educational programming for students and budding innovators to learn from experts in the field. This video collection includes descriptive terminology and written descriptions of the videos and is discoverable via EHSL's e-channel.

In the final chapter, the future of innovation and information is explored, and insights are shared about how these two *I* words have so much to gain when they collide. As innovators need to understand the contexts within which they are designing new products, games, technologies, and processes, they need to build on existing knowledge to gain new insights and perspectives that guide their product or process innovation. They are dependent on past literature to provide this knowledge and foundation and thus benefit from partnering with librarians who know how to discover relevant evidence using their extensive training and expertise. As universities become more dependent on innovation as an income source, how will this dependency impact libraries and librarians and their futures? Several forecasts of how libraries will be shaped by their partnership with innovators are offered. The future is unknown, but librarians can prepare by understanding more about innovation and the knowledge needed by innovators in order to be effective and productive partners.

Through reading this book, you will gain an overview of how important information is to innovation and, in return, how important innovation is to information and librarians' futures. See the following list of websites, which are frequently referenced throughout this book. Enjoy the book and contact the editors with your comments and thoughts after reading; we are interested in hearing from you!

LIST OF COMMONLY USED WEBSITES

B-2-B	http://cmi.uofuhealthsciences.org/bench-2-bedside
Bench to Bedside	http://cmi.uofuhealthsciences.org/bench-2-bedside
BioImmersion	http://design.cap.utah.edu/2016/03/bioimmersion-program-opportunity
BioInnovate	http://www.bioinnovate.utah.edu
Center for Medical Innovation	http://cmi.uofuhealthsciences.org
CMI	http://cmi.uofuhealthsciences.org
DiGRA	http://www.digra.org
Digital Games Research Association	http://www.digra.org
e-channel	http://library.med.utah.edu/e-channel
EAE	http://eae.utah.edu
Ecosystem	http://www.utah.edu/innovate
EHSL	http://library.med.utah.edu
Entertainment Arts and Engineering	http://eae.utah.edu
Entrepreneurial Faculty Scholars	http://www.efs.utah.edu
James E. Faust Law Library	http://www.law.utah.edu/library
Games4Health	http://g4h.business.utah.edu
I-Corps	http://cmi.uofuhealthsciences.org/i-corps

Ideation Studio	http://library.med.utah.edu/esynapse/2016-july-september-vol-31/ideation-studio-opens
Innovation Ecosystem	http://www.utah.edu/innovate
Innovation Life Cycle	http://library.med.utah.edu/e-channel/portfolio/innovation-life-cycle
Innovation Vault	http://library.med.utah.edu/e-channel/portfolio/the-innovation-vault
J. Willard Marriott Library	http://www.lib.utah.edu
Lassonde Entrepreneur Institute	http://lassonde.utah.edu
Lassonde Studios	http://lassonde.utah.edu/studios
Medical Library Association	http://www.mlanet.org
MLA	http://www.mlanet.org
School of Business, University of Utah	http://eccles.utah.edu
Sorenson Center for Discovery and Innovation	http://digitalsandboxu.com
Spencer S. Eccles Health Sciences Library	http://library.med.utah.edu
Technology and Venture Commercialization	http://www.tvc.utah.edu
The GApp Lab	http://thegapp.eae.utah.edu
Therapeutic Games and Apps Lab	http://thegapp.eae.utah.edu
TVC	http://www.tvc.utah.edu
U-Bar	http://healthcare.utah.edu/hospital/library/u-bar-apps.php
University of Utah	http://www.utah.edu
University of Utah Bench to Bedside	http://cmi.uofuhealthsciences.org/bench-2-bedside

University of Utah BioImmersion	http://design.cap.utah.edu/2016/03/bioimmersion-program-opportunity
University of Utah BioInnovate	http://www.bioinnovate.utah.edu
University of Utah Center for Medical Innovation	http://cmi.uofuhealthsciences.org
University of Utah David Eccles School of Business	http://eccles.utah.edu
University of Utah e-channel	http://library.med.utah.edu/e-channel
University of Utah Ecosystem	http://www.utah.edu/innovate
University of Utah Entertainment Arts and Engineering	http://eae.utah.edu
University of Utah Entrepreneurial Faculty Scholars	http://www.efs.utah.edu
University of Utah Games4Health	http://g4h.business.utah.edu
University of Utah Health Care	http://healthcare.utah.edu
University of Utah Health Sciences	http://healthsciences.utah.edu
University of Utah Ideation Studio	http://library.med.utah.edu/esynapse/2016-july-september-vol-31/ideation-studio-opens
University of Utah Innovation Ecosystem	http://www.utah.edu/innovate
University of Utah Innovation Vault	http://library.med.utah.edu/e-channel/portfolio/the-innovation-vault
University of Utah J. Willard Marriott Library	http://www.lib.utah.edu

University of Utah James E. Faust Law Library	http://www.law.utah.edu/library
University of Utah Lassonde Entrepreneur Institute	http://lassonde.utah.edu
University of Utah Lassonde Studios	http://lassonde.utah.edu/studios
University of Utah School of Business	http://eccles.utah.edu
University of Utah Spencer S. Eccles Health Sciences Library	http://library.med.utah.edu
University of Utah Technology and Venture Commercialization	http://www.tvc.utah.edu
University of Utah The GApp Lab	http://thegapp.eae.utah.edu
University of Utah Therapeutic Games and Apps Lab	http://thegapp.eae.utah.edu
University of Utah TVC	http://www.tvc.utah.edu
University of Utah U-Bar	http://healthcare.utah.edu/hospital/library/u-bar-apps.php
VentureWell	https://venturewell.org

NOTES

1. Todd J. Brinton, Christine Q. Kurihara, David B. Camarillo, et al., "Outcomes from a Postgraduate Biomedical Technology Innovation Training Program: The First 12 Years of Stanford Biodesign," *Annals of Biomedical Engineering* 41, no. 9 (2013): 1803–10, doi:10.1007/s10439-013-0761-2; Mary R. Goldberg and Jonathan L. Pearlman, "Best Practices for Team-Based Assistive Technology Design Courses," *Annals of Biomedical Engineering* 41, no. 9 (2013): 1880–88, doi:10.1007/s10439-013-0798-2; Jaspal S. Sandhu, Robert N. Hosang, and Kristen A. Madsen, "Solutions That Stick: Activating Cross-Disciplinary Collaboration in a Graduate-Level Public Health Innovations Course at the University of California, Berkeley," *American Journal of Public Health* 105 (suppl. 1, 2015): S73–77, doi:10.2105/

AJPH.2014.302395; Youseph Yazdi and Soumyadipta Acharya, "A New Model for Graduate Education and Innovation in Medical Technology," *Annals of Biomedical Engineering* 41, no. 9 (2013): 1822–33, doi:10.1007/s10439-013-0869-4.

2. Patrick D. Loftus, Craig T. Elder, Troy D'Ambrosio, and John T. Langell, "Addressing Challenges of Training a New Generation of Clinician-Innovators through an Interdisciplinary Medical Technology Design Program: Bench-to-Bedside," *Clinical and Translational Medicine* 4 (2015): 15, doi:10.1186/s40169-015-0056-3; Patrick D. Loftus, Craig T. Elder, Matthew W. Sorensen, et al., "Creating a Benchmark Medical Technology Entrepreneurship Competition—The University of Utah Bench-to-Bedside Medical Device Design Competition," *NCIIA OPEN 2014 Conference Report*, 2014, http://venturewell.org/open2014/presentation/creating-a-benchmark-medical-technology-entrepreneurship-competition (accessed November 20, 2016).

Chapter One

Deep History

Creativity, Innovation, and Libraries

Joseph Lucia

An opening question: when we focus on change, do we lose sight of what endures? For what endures forms the heart of the "long now" of any abiding institution.[1] In looking back, continuity of purpose can be just as striking as disjunction and historical difference.

In an environment where the impact of the digital technology revolution confronts us daily and where notions of business and institutional disruption are mostly embraced as intrinsically valuable, it is useful to telescope our perspective on the present to gain a stronger purchase on the deep history undergirding libraries and their kindred enterprises. Unearthing that history enables us to fashion a template for our continuing vitality in a networked digital world.

Consider the following description from Lionel Casson's seminal study *Libraries in the Ancient World* on the establishment of Alexandria as a great cultural center:

> What helped mightily in enticing intellectuals to the city was the founding by Ptolemy I of the famous Museum. In ancient times, the word museum normally referred to a religious establishment, a temple for the worship of the muses; Ptolemy's creation was a figurative temple for the muses, a place for cultivating the arts they symbolized.[2]

Casson further characterizes the museum as "an ancient version of a think tank: the members, consisting of noted writers, poets, scientists, scholars."[3] Provided room, board, and a salary, he adds, the members of the museum "had at their disposal a priceless intellectual resource: it was for them that the Ptolemies founded the library of Alexandria."[4]

Reflecting on this same historical context, Schnapp and Battles write, "The world-making properties of the library—the *theca* as microcosm—are enduring facets of its container function."[5] They extend this argument to claim that these ancient libraries rest on a founding paradox:

> On the one hand, they are places of enclosure: fortified bastions; sites of burial and storage of treasures; places of retreat from the din of the market place; sacred precincts and temples devoted to contemplation and prayer; self-sufficient worlds where, like their monastic descendants 10 centuries later, a community of insiders, the priest and his retinue of *philologoi*, or lovers of the word, hold court. On the other hand, libraries open up *onto* the world: the noise of the street invades the sacred precinct; their collections cannot be built up without connections between capital and periphery in the form of trade routes; connections between society at large and communities of learning.[6]

This notion of the library as situated on the fulcrum of two distinct domains informs contemporary thinking about the nature and value of library physical spaces. Craig Dykers, founding partner of the design firm Snøhetta, designers of the new Library of Alexandria, and also architects for the new library building at Temple University in Philadelphia, wrote the following e-mail (November 6, 2014) to the author about the liminal spaces connecting ancient library complexes to their larger urban settings:

> It would not be uncommon, in the same way as it does today, that the world of knowledge would leak out of the formal institutions and be confronted in more public places far away from the classrooms. Although the gymnasia were somewhat informal, they seem to have played a defined role in society. Outside of these places there were markets, taverns, streets and so on where one would imagine the philosophers and academics would meet, sometimes informally.
>
> One such place that is fascinating to me is the *stoa*, a portico that is something like a mix between a marketplace, an art gallery, and a community hall. I believe the *stoa* were very lively places and they often seem to be located centrally in communities, near the main markets. There would be wine, food, fragrance, light, and air to nurture the senses.

The restored Stoa of Attalos in Athens, depicted in figure 1.1, is a fine example of such a space. Dykers goes on to liken the experience of walking through the *stoa* to traversing the carefully planted groves of trees that defined the boundaries of the earliest academies in the ancient world (hence the phrase *the groves of academe*):

> I believe that the dimensions and design of the *stoa* were meant to encourage interaction between people. And now after learning about the Academus grove

I can even see the embodiment of the trees in the design of the columns, but that is perhaps just me. I see these more as trees because you walk alongside them more than you walk through them, in the same way you would walk along a planted row of plane or olive trees. The variation of dark and light from one side to the other also helps us gauge our movement. All of this leads to the same kinds of features that sport activities lead to and that is a greater awareness of yourself and your context, sharpening the mind. And what is a library without a mind to inhabit it?

Figure 1.1. Stoa of Attalos in Athens. *Wikimedia Commons, CC BY permissions.*

How, then, can we get from classical architecture to a more current sense of the evolving function of the academic library in particular? In the recent past, this linkage of libraries to sites of inspiration has taken on new resonance. Sam Demas, in his superb contribution to a 2005 collection of essays on *library as place*, provides the following formulation:

> The design of public and academic libraries is beginning to embody an egalitarian renaissance of the ideal of the *Mouseion* at Alexandria. Generally remembered as the Library of Alexandria, the *Mouseion* was indeed a great synoptic collection. However, its larger purpose is lost from popular memory and is indeed largely missing from our conception of the library in higher education today. The "temple of the muses" was a research center, a museum, and a venue for celebrating the arts, inquiry, and scholarship.[7]

Demas then suggests that "the college library look to the *Mouseion* as one model for further integrating itself into the community it serves and for providing a unique cultural center that inspires, supports, and contextualizes its users' engagement with scholarship."[8]

Seen in this light, the library becomes both a physical and a conceptual nexus between the various disciplines and domains that embody the intellectual heterogeneity of a college or university environment: crossroads, commons, public square, and contact zone. This is a uniquely valuable social space, one rich with potential for fostering the collisions and chance interactions among those from different backgrounds and distinctive perspectives.

Steven Johnson, in his book *Where Good Ideas Come From: The Natural History of Innovation*, makes a compelling case for the view that the generation of novelty occurs in those zones where different environments, different viewpoints and frames of reference, intersect and overlap.[9] His argument pertains equally to the generation of biodiversity in nature as it does to the creation of new ideas within human communities and networks of information exchange. Johnson's study of conditions for innovation culminates in a classification scheme that breaks out instances of creative and scientific breakthrough into four quadrants: (1) market/individual, (2) nonmarket/individual, (3) market/networked, and (4) nonmarket/networked. The quadrants distinguish between individually motived versus group-generated (e.g., network-generated) innovation and invention and between innovation and invention motivated by commercial/market opportunities versus innovation and invention triggered by curiosity and playful inquiry without a primary focus on marketplace opportunities or the production of new *products* in a purely commercial sense. His goal in exploring that typology of innovation events is to determine how commercial versus noncommercial and individual versus group action contribute in aggregate over time to the total number of significant intellectual, technical, and cultural breakthroughs and inventions. What he finds, just as a matter of counting significant innovations over time, is that the nonmarket/networked quadrant is the most productive by an impressive margin. He concludes his analysis by saying,

> Much of the history of intellectual achievement over these past centuries has lived in a less formal space [than decentralized markets and command-and-control states]: in the grad seminar and the coffeehouse and the hobbyist's home lab and the digital bulletin board. The fourth quadrant should be a reminder that more than one formula exists for innovation. The wonders of modern life did not emerge exclusively from the proprietary clash between private firms. They also emerged from open networks.[10]

The touchstone concept here is "openness"—and behind that the construct of the *commons* and, more specifically, of the *knowledge commons* as a space for equal and open (communal) access to ideas without proprietary boundaries. Libraries function implicitly for their communities as a knowledge and cultural commons: the intellectual and creative goods in a library's physical and digital collections are made freely available to anyone in the community who needs or desires them. The underlying principle of that openness is that the unimpeded circulation of ideas is an implicit good, one that in effect supports the creation of new knowledge over time. Nancy Kranich has made this case with specific reference to the evolving role of academic libraries in her essay "Countering Enclosure: Reclaiming the Knowledge Commons," where she writes, "As information distribution becomes more diffused, libraries become more involved in the process of scholarly communication and in building information communities. This transformation into more engaged, collaborative institutions will transform libraries as creators and not just sustainers of knowledge commons."[11]

From this follows a key assumption, one that implicitly lodges creativity and innovation at the core of the library mission: the unique societal role of libraries is to support, expand, and preserve the intellectual and cultural commons in all of its discursive and literary forms and, through that commitment, to nurture the generative conversation begetting new knowledge and new cultural expression.

Both through the resources they avail to their communities and through their provision of physical settings for community (interdisciplinary) engagement, libraries exist to catalyze creativity and innovation. Always more than warehouses, libraries are, to use the words of Schnapp and Battles, "a kind of machine—an engine for learning, provocation, and discovery—[that] can take us many places."[12]

When we talk about creativity, innovation, and the generation of new ideas, it is almost a matter of ritual practice to invoke *inspiration* as an essential ingredient. Again here, there is a way this fits into the social and cultural terrain occupied by libraries, specifically in relation to the architectural character of the most sublime library buildings, and the way they embody a particular aspiration for the role of libraries in the lives of their users. Libraries are, at least as much as any other building type, deeply imbued with metaphors that shape and express the experiences of their users. Architect Brook Muller has written of the great Louis Kahn Library at Philips Exeter Academy in New Hampshire that it "furnishes a compelling example of a library project that derives meaning through correspondences between the metaphorical, spatial, and experiential, organized as it is around the notion of taking a book and bringing it to the light."[13]

Elaborating on this point, Muller goes on to quote himself from an earlier essay:

> A conceptual notion (how we gain knowledge) originates in embodied experience (vision made possible by the presence of light, such that light enabling

vision = knowledge). Kahn projects this construct back upon the physical realm through patterns of spatio-luminous organization "embodied" in the Library: one literally takes a book from a low ceilinged and relatively dark "stack space" and brings it to a generously daylit study carrel at the building's periphery.[14]

Let us look briefly at the way library spaces speak to and shape experience in three particularly dramatic (and perhaps also paradigmatic) examples that appear designed to engage and inspire.

The first is the famous 1857 British Museum Reading Room in figure 1.2. Based on the design of the Roman Pantheon, this remarkable domed space situated its occupants at the center of the universe of knowledge and invited them to participate in the act of reading and reflection among a community of quietly busy scholars. It is clearly much more than a purely functional design, perhaps even overtly imperialist in its appropriation of a classical form as a call to England's intellectual class to aspire to world-historical greatness. It is fitting to mention on a somewhat subversive note that Karl Marx did much of his economic and social research in that very room.

Figure 1.2.   1857 British Museum Reading Room. *Wikimedia Commons, GNU Free Documentation License.*

The second is the Graduate Reading Room at the Suzallo Library of the University of Washington (figure 1.3). This great space dates from 1926 and represents one of the pinnacles of collegiate gothic. Here, the ecclesiastical impulses of the design imply an awesome solemnity to the act of study through a soaring, arched ceiling; tall tracery-covered windows; and a transparent nave at the far end of the room. In this place, the encounter with knowledge and the act of study is implicitly coextensive with spiritual growth and enlightenment.

The third is more contemporary, a public seating/reading area in the main branch of the Seattle Public Library, which opened in 2004 (figure 1.4). Again, we are looking at a dramatic, vaulted space that adheres to the metaphor that Brook Muller found in the design of the Kahn Library discussed above—the incursion of light from the steeply canted exterior wall washing

Figure 1.3. Graduate Reading Room, Suzallo Library, University of Washington.
Wikipedia, CC BY-SA 3.0 permissions.

over readers in the adjacent seating areas and connecting the life of thought and reflection to the vibrancy of the outside urban world. Although slightly more hermetic, this space is in some manner akin to the *stoas* lauded by Craig Dykers above. Thinking along more contemporary lines, Dykers offered some additional thoughts about libraries in an e-mail to the author (March 23, 2016) in preparation for a panel about libraries and cities in the 21st century. He wrote,

> Being an architect I believe architectural issues play an important role in the creation of a library, or any typology. But it is not appearance or style that interests me as much as geometry, light, movement, air, vegetation, and sustenance. These were the same things that the ancient libraries and academic institutions were built upon. Of course ornament, color, massing are important but they are not as primary as these essential qualities.

And yet again, we are in the realm of metaphor: light, enlightenment, circulation, and sustenance—the use of formal properties to express the social function of a library environment.

In that same communication, Dykers added,

> Italo Calvino postulated in *Invisible Cities* that we all expect or imagine cities as being occupied by certain things, we have certain expectations of what a city must be. We feel we need libraries in our cities, even if we do not use them. But the truth is that libraries are vital and used regularly by many people today, so they do not need to be imagined. Libraries not only provide valuable space for research, entertainment, and development, they also are supporting many people who have few options to access knowledge sources. They are as

Figure 1.4. Reading Area, Main Branch of Seattle Public Library. *Wikipedia, CC BY 2.5 permissions.*

much about making as taking today. In the future libraries will not just be about research as they were in the past, they will be where books are written and printed, where objects are conceived and produced, where music is made and not simply checked out.

This brings us back, then, to the central matter here: the abiding but also evolving place of creativity, innovation, and *making* in the context of library mission and role. The recent makerspace movement is in some ways a re-awakening of a quiet (one might say hidden) long-term function of library environments as settings for knowledge creation. 3D printers, laser cutters, and electronics kits may be a fashion of the moment, but they reflect an implicit conviction that the *stuff* libraries provide their communities is as much the stuff of making as it is the stuff of recorded thought and culture. Jim O'Donnell, university librarian at Arizona State University and former provost of Georgetown University, speaking at the Charleston Conference in 2015, made the following pertinent observation:

> We no longer should think of libraries, we should no longer think of knowledge, as stable collections of information which can be consulted and used and recorded and preserved. Rather, we need to think of our libraries as places in which new knowledge is being made all the time. Many of us have been experimenting with so-called maker spaces in our libraries. My argument here is simply that we have always been a maker space but now we need to conceive ourselves in those terms, conceive the support of users as people who are making knowledge all the time making new knowledge using our tools.[15]

Let us take this one small step further. The move we are making in libraries at this moment is to render a quietly implicit function of our *long now* into a deliberately amplified and explicit dimension of our 21st-century mission in an era *beyond the book*: a shift in the center of gravity rather than any sort of alteration of the grounding principles and functions of libraries. We make the commons a vital source for the creation of new ideas, new knowledge, and new cultural expression. That requires that we bring to our work a greater consciousness of the continuum between inquiry and discovery, reading, reflection, and creation. In libraries, our work is now as much about framing, production, expression, and dissemination as it is about storing and organizing information. This new reality is expressed in many ways, but some of the more obvious come directly out of evolving professional practice, including the creation of digital libraries from our Special Collections; the support for open-access and scholarly publishing initiatives; the development of research data management services; the founding of digital scholarship centers and digital humanities programs; the growing and deepening connections between many academic libraries and their associated university presses; the aforementioned establishment of makerspaces, innovation labs, and start-up incubators in public and academic libraries; the increasing adoption of open-technology and open-source software solutions to support more user responsive online environments (including a burgeoning cohort of technology development talent in many larger libraries); and multi-type library collaboration in building the Digital Public Library of America (http://dp.la).

David Lankes, noted iSchool and library science faculty member at Syracuse University and now at the University of South Carolina, has devoted much of his career around rethinking librarianship as an engaged *making-oriented* profession. The mission statement for the *new librarianship* (and hence of *new librarians*) is "to improve society through facilitating knowledge creation in their communities."[16] At my institution, as we have programmed and designed a new library building with our architects, we settled on a mission statement for the facility that articulates this explicitly. That mission says,

> A catalyst for learning and intellectual engagement, the new Temple University Library cultivates and supports scholars and the scholarly enterprise, connecting people and ideas in a compelling resource and service-driven environment. Calling its diverse communities together for inquiry and exploration, this evocative building inspires the discovery, creation, preservation, and sharing of knowledge.[17]

We have an almost unprecedented opportunity at this moment to build on and intensify the buried legacy of making and innovation that is at the historical root of the institution of the library. We might even view this as a chance to reinvent librarianship as a *creative class* profession. That assertion makes some of the more staid among our professional ranks nervous and even uncomfortable. Perhaps we should see those anxieties as indicators that though it feels as if we are stepping into the unknown, we are in fact on the right track. For the future of libraries, we must make it the case that creativity and innovation are woven into our sense of purpose, are part of what endures.

NOTES

1. The Long Now Foundation "About Long Now," http://longnow.org/about (accessed November 12, 2016). The Long Now Foundation "hopes to provide a counterpoint to today's accelerating culture and help make long-term thinking more common . . . to foster responsibility in the framework of the next 10,000 years." It's worth noting that the foundation has as one of its projects a "library of the deep future, for the deep future. In a sense every library is part of the 10,000-year Library, so Long Now is developing tools (such as the Rosetta Disk, The Long Viewer and the Long Server) that may provide inspiration and utility to the whole community of librarians and archivists."

2. Lionel Casson, *Libraries in the Ancient World* (New Haven, CT: Yale University Press, 2002), 33.

3. Ibid.

4. Ibid., 34.

5. Jeffrey T. Schnapp and Matthew Battles, *The Library beyond the Book* (Cambridge, MA: Harvard University Press, 2014), 26–27.

6. Ibid.

7. Sam Demas, "From the Ashes of Alexandria: What's Happening in the College Library," in *Library as Place: Rethinking Roles, Rethinking Space*, ed. Scott Bennett, Sam Demas, Geoffrey T. Freeman, Bernard Frischer, Kathleen Birr Oliver, and Christina A. Peterson, CLIR Publication *129* (Washington, DC: CLIR, 2016). Accessed November 6, 2016. http://www.clir.org/pubs/reports/pub129.

8. Ibid.

9. Steven Johnson, *Where Good Ideas Come From: the Natural History of Innovation* (New York: Riverhead Books, 2010).

10. Ibid., 236.

11. Nancy Kranich, "Countering Enclosure: Reclaiming the Knowledge Commons," in *Understanding Knowledge as a Commons: From Theory to Practice*, ed. Charlotte Hess and Elinor Ostrom (Cambridge, MA: MIT Press, 2007), 104.

12. Schnapp and Battles, *The Library beyond the Book*, 91.

13. Brook Muller, "Cabinet, Vault and Luminous Forest: Complexity and Contradiction in Green Library Architecture" (paper presented at the Association of College and Research Libraries Meeting, Portland, OR, March 28, 2005), http://www.ala.org/acrl/sites/ala.org.acrl/files/content/conferences/confsandpreconfs/2015/Muller.pdf.

14. Brook Muller, *Ecology and the Architectural Imagination* (New York: Routledge, 2014), 59.

15. Jim O'Donnell, "Star Wars in the Library," transcribed from a presentation given at the 2015 Charleston Library Conference, Charleston, SC, November 4–7, 2015, https://www.youtube.com/watch?v=O0aCUUNxRpA&feature=youtu.be.

16. David Lankes, "New Librarianship," http://davidlankes.org/?pageid=6352 (accessed October 4, 2016). His book *The Atlas of New Librarianship* (Cambridge, MA: MIT Press, 2011) is a touchstone text for anyone wishing to understand how the work of knowledge creation connects to our professional core competencies.

17. Temple University, *New Library Design Brief*, 12 (prepared in collaboration with brightspot/Snohetta, Philadelphia, 2014).

Chapter Two

Innovation Cycle and Information Applications

Jean P. Shipman, Tallie Casucci, and Spencer W. Walker

Innovators commonly proceed through four main stages when creating their technological products or processes, henceforth called the technology innovation life cycle. These four main stages are (1) ideation and concept development, (2) prototyping, (3) product, and (4) commercialization. These stages require different degrees of focus over the course of development and various kinds of knowledge, but almost all of the stages require different kinds and levels of information. Progress through these stages is not linear in practice and may occur in a different order than outlined, and some stages may need to be repeated. This chapter reviews these technology innovation stages and elaborates on how the University of Utah (U of U), including the Spencer S. Eccles Health Sciences Library (EHSL) and other libraries at the U of U, offer assistance at each stage of the innovation process to provide examples for other universities to emulate or improve upon.

STAGE 1: IDEATION AND CONCEPT DEVELOPMENT

Ideation happens when a concept starts to originate within an innovator's mind-set and is given consideration as to its viability, its value proposition, and how it solves a problem. Identifying a true problem and its root cause is part of this stage of the innovation process. Deriving a solution to the problem can take quite a lot of exploration. It is important to review the literature to both understand the problem and determine if the problem has already been adequately addressed and solved by others. Identifying prior art through patent searching is one way to investigate how others have attempted to solve the problem and to see what solutions have been claimed via patent law. Through patent searching, one can identify potential competitors, legal experts within the field, and if the idea is novel, thus patentable. Taking courses offered at universities and in the community is another way to understand how to initiate this first step of the innovation process. Reading about the product topic is another way to learn. Moreover, it is important to query the potential users of the product and ask questions such as, "Does my solution actually meet a real need?" and "What do health professionals and patients really want?"

Brainstorming with others, particularly with those experienced in the area of interest, is a great way to consider different aspects of an issue or new product. Having a diverse group of individuals with different backgrounds lends a variety of perspectives and approaches to deriving solutions. IDEO, a design and innovation consulting firm, developed rules and best practices for innovative brainstorming (http://www.ideou.com/pages/brainstorming).

Mentors can also be a great way to engage with others to test out your ideas and to validate the need for any product under consideration. Mentors can be entrepreneurs, faculty, other students, or anyone with experience with innovation. Mentors can be local or global and can be used circumstantially—for specific aspects—or can guide throughout the entire development process. Mentors can be self-identified or identified and referred by others. Mentors can be formal or informal and can be invaluable, as they often have the wisdom and experience needed to help shape new product designs.

To help guide one through the entire innovation life cycle, Osterwalder and Pigneur have created a Business Model Canvas[1] (BMC) that is a template with nine steps to follow. Five of the nine BMC steps correlate to the ideation stage: value proposition, key activities, key partners, key resources, and customer segments. More information about the BMC is available on e-channel (http://library.med.utah.edu/e-channel/innovation-vault-business-model-innovation; see chapter 13). Understanding what customer needs a product will address, identifying who should be involved in product development and needed resources for its creation, and determining how funding for product development will be obtained are examples of product ideation discussions. Identifying the targeted audiences for a product is also a component of the ideation stage of innovation.

U OF U OFFERINGS—STAGE 1

At the U of U, multiple offerings help innovators with the ideation and concept development stage of the innovation life cycle. These offerings are explained below.

Instruction

At the U of U, there are several academic and certificate courses as well as programs related to innovation. BioInnovate is a yearlong course offered by the U of U Department of Bioengineering for graduate students to observe clinical situations, identify visible problems, and work with faculty to develop productive solutions. BioImmersion is a similar course for undergraduate students offered during the summer. The Lassonde Entrepreneur Institute offers courses requiring a smaller time commitment, including a three-credit course to anyone wishing to learn more about entrepreneurship. The David Eccles School of Business offers students major, minor, PhD degrees, and a certificate program as well.

Space

The EHSL offers an Ideation Studio space that is equipped with carts of widgets and gadgets to conduct group discussions and brainstorming sessions. Several telescopic movable whiteboard dividers accommodate any size group and any shape of space needed. Any U of U individual can reserve the space, and there is no charge for the supplies. Instructional sessions on basic design principles, human factors, and Design Box (chapter 12) applications are taught to interested parties.

The EHSL offers another space, the Synapse (chapter 3), which is located on its lowest level, the Garden Level. It includes meeting rooms, conference rooms, and collaborative spaces divided by mobile telescopic whiteboards complete with large display monitors. There is also a small café and kitchenette space that encourages informal group interchanges and exploration of problems posed by a Twitter hashtag—#asksolve (http://library.med.utah.edu/blog/eccles/2016/01/19/asksolve). Individuals can have a cup of coffee, eat lunch and talk about the submitted problems, or congregate on demand to address designs, develop and refine product or process concepts, and plan strategies of attack for moving to the next innovation stage.

Recommended Readings

As with most topics, one can find a large cadre of books and journal articles about the art of ideation. At the U of U, several books are available in a LibGuide called *Innovation Guide*.[2] Entrepreneur and design faculty members helped to identify content for the *Innovation Guide*. This collection includes the aforementioned book about the BMC and is maintained by the Libraries Innovation Team (chapter 10).

BMC

The EHSL has produced large circulating vinyl hanging versions of the BMC that teams can use when developing their product infrastructure and business and marketing plans. Teams can use sticky notes with these templates to complete the nine steps in the BMC. In addition, the EHSL has taken the BMC and identified information resources and databases that assist with completing the template steps.[3] This unique outline of matching resources per BMC step provides a type of *cheat sheet* to innovators wishing to research and validate the need for their product.[4]

Mentors

The U of U has developed several ways to identify potential mentors. There is a website that connects researchers to one another called *Colleague*, developed by The GApp Lab in conjunction with the U of U's Health System Innovation and Research Division of the Department of Population Sciences, School of Medicine.[5] There is also a group called Entrepreneurial Faculty Scholars, which encourages faculty entrepreneurs at the U of U to share their expertise and to assist others in the innovation process. Several industry members and alumni have offered their assistance to specific competitions and programs. For example, the Bench to Bedside (B-2-B) competition (chapter 10) includes a *meet your mentor* event where teams can meet the various mentors.

Patent Searching

As mentioned earlier, patent searching identifies prior art and ensures that an innovation is indeed novel or innovative. There are several librarians at the EHSL and the J. Willard Marriott Library (main academic U of U library) who are part of a Libraries Innovation Team (chapter 10) that guide innovators to use the U.S. Patent and Trademark Office's resources, including its patent search database. The patent librarian teaches numerous workshops and maintains the patent LibGuide.[6] There are also legal fellows employed as part of the Center for Medical Innovation (CMI) who provide consultations to B-2-B teams about how to apply for provisional patents (http://www.uspto.gov/patents-getting-started/patent-basics/types-patent-applications/provisional-applica
tion-patent) and how to complete required registration paperwork.

STAGE 2: PROTOTYPING

Once a product concept has been conceived, investigated, and deemed viable for development, the next innovation stage is to create a mock-up of the product to test its specifications and its ability to solve the problem under consideration and to refine the design and fit for the targeted audience. This is termed *prototyping*. Since this chapter addresses mainly technological innovations, the prototypes are usually physical models of the product that can be held, manipulated, and studied in greater detail. Again, interprofessional teams of experts, including an engineer, should consult on the prototype design.

U OF U OFFERINGS—STAGE 2

The U of U offers several laboratories for creating prototypes of devices and products. The health sciences–related laboratory is located in the EHSL and is called the Gary L. Crocker Innovation and Discovery Laboratory (chapter 6). This fabrication lab has an extensive list of equipment, including 3D printers, and it provides additive, subtractive, and electrical prototype developments with expertise offered by staff, fellows, and mentors. This lab is open to all U of U personnel. The B-2-B teams use their prototyping funds to pay for lab material costs and staff expertise.

The Marriott Library has a variety of material samples available for touching and feeling in its Materials Collection, which includes the *Materials ConneXion* (MC) database (http://campusguides.lib.utah.edu/MCXUtah). This searchable database includes more than 7,000 materials and design processes in a variety of fields of design. One can search by material category (e.g., glass, polymers, and metals), by processing types (e.g., welding, lamination, and injection molding), and by sustainability, usage, manufacturer, and physical properties. It is also possible to search by an MC material inventory number, if known.

Within the new Lassonde Studios, home for more than 400 U of U students interested in innovation, a 20,000-square-foot garage space or "creative workshop" is open 24/7 for resident students to prototype their devices and concepts. Mentors are also available, including Lassonde student ambassadors, to help teams further shape their ideas and designs.

Other fabrication laboratories are available within the U of U Innovation Ecosystem, including a Utah Nanofabrication Laboratory and other facilities within the College of Engineering's Center for Engineering Innovation. There is also a Materials Characterization Lab located in the U of U Department of Materials Science and Engineering that offers a wide range of analytical instrumentation and services, such as a scanning electron microscope, an X-ray diffractometer, and a tabletop Instron (5969).

For U of U personnel seeking funding to assist with prototype development, there are several sources available for students, including B-2-B funding and monthly *Get Seeded* opportunity funding by the Lassonde Entrepreneur Institution. For faculty, the U of U is a National Science Foundation (NSF) I-Corps host site; $3,000 is given to project teams to test their innovative ideas, with the idea of going for a large NSF grant after seed funding. The funding supports educational sessions as well as access to the *LaunchPad Central* software (https://www.launchpadcentral.com).

STAGE 3: PRODUCT

Creating a prototype of a product, as described in stage 2, facilitates final product design, which is stage 3 of the innovation life cycle. A prototype offers a physical demonstration of the product concept or idea and the ability to test to see if the product operates or looks like it was intended. A few issues to consider at the product stage of development are the following: Have the right materials been selected for the conditions under which the product will be primarily used? Are the materials and design safe and effective for the end user? Does the design consider general human factors (work well with users' hands or feet or whatever part of the body interacts with the product)? What is the manufacturability of the product? Can it be mass-produced easily and at an efficient cost point? Is there a material that could be used that would make the production cost less without reducing the product's safety and effectiveness? Are there design alterations that need to be made to increase the design effectiveness? Is the product design attractive? Are there refinements that can be made that would improve the appearance, lessen the weight, reduce the size, or lower the cost?

The product stage also includes intensive testing of the created device or product. If it is a medical device, clinical trials might be required to attest to the safety and effectiveness of the product and to confirm that it performs as its specifications stipulate. Having potential users test the product with a battery of questions for them to address is a great way to ascertain that the specifications do meet the desired need and application. In the United States, this stage of innovation may also require approval from federal agencies, depending on the product type and classification. The Food and Drug Administration (FDA) has several processes and regulations that a product is subjected to in order to affirm that it is safe and effective for human use. There are a plethora of standards, nationally and internationally, that may be required to be met. The standards organizations include the American National Standards Institute, the American Society for Testing and Materials, and the International Organization for Standardization (ISO). There are several regulatory- and quality-related hoops to jump through in order to get a product to the market. The use of standards helps give direction and guidance to ensure that the product is designed, tested, and manufactured in a way to get through those hoops.

Once the product is designed and meets the quality requirements, including safety and effectiveness, an innovator needs to determine the manufacturability of the product. Possible qualified suppliers must be identified, along with details such as where are they located and if they are willing to commit to being a supplier based on terms established, including standard quality agreements. Another item to refine further at this time in the innovation process is the final business model for the product. In other words, what is the cost to create or produce the product, and what will be the desirable and effective manufacturing mechanism?

Referring again to the BMC, the final four steps of the nine-part template apply to this third stage of innovation. These steps include the customer relationships, cost structure, key resources, and key partners. These steps outline how to identify suppliers, manufacturers, and distributors in order to be able to commercialize a product. They help outline what resources are needed to manufacture and mass-produce a product and determine associated costs. Cost itemization is important, as it will help to determine whether the manufacturing process is cost effective enough and in line with the desired price point for the product, which is needed for the next stage of innovation—commercialization. Customer relationships are critical to product success in that they determine the level and effectiveness of agreements reached with suppliers, manufacturers, and distributors.

U OF U OFFERINGS—STAGE 3

The EHSL has provided support to medical device innovators by purchasing a series of ISO standards on biological evaluation of medical devices (ISO 10993), medical device quality management systems (ISO 13485), electrical equipment (IEC 60601), and risk management (ISO 14971). These ISO standards are in EHSL's online catalog and are located within the Gary L. Crocker Innovation and Design Laboratory (chapter 6). Additionally, the Marriott Library has gathered information related to ISO and other standards in a LibGuide called the *Industry Standards Guide*.[7]

The CMI has hired a full-time staff member who has extensive knowledge and working experience with the various FDA and international regulations and who can assist innovators with establishing their design, development, and manufacturing requirements, including completing the variety of paperwork required to meet regulatory reviews, quality auditor inspections, and manufacturing approvals. This individual can also connect innovators to the clinical trials personnel within the U of U Office of Research to be able to learn about clinical trial policies and procedures.

The Technology and Venture Commercialization (TVC) office is the tech transfer office at the U of U. It assists innovators with identifying potential suppliers and manufacturers and advises on business plans and market impact. The TVC has expertise in all areas of business plan development and can draw on its almost 40 years of experience to guide new innovators through the various processes required to take a product to market. The TVC also has staff that provide direction on filing provisional patents and has access to legal counsel for such filings.

STAGE 4: COMMERCIALIZATION

The final innovation stage deals with taking a product to the market and determining a price point, buyers, and how to distribute the product via various channels. It includes developing ways to meet commercial demand for the product and to market it to encourage such demand. Business plans and promotional plans are key components of this stage. Who are the customers? How will they learn about the product? What will be the selling features to highlight in promotions? What price should be charged for the product? All of these questions need to be considered prior to releasing a product.

Again, the BMC comes to the rescue, as it helps one revisit answers to these questions with research and verification techniques, such as customer interviews and a determination of appropriate costs, margins desired, and identified customer distribution outlets. The steps of the BMC that apply to this fourth stage include customer relations, cost structure, channels, revenue streams, and customer segments. There are many resources available on the Internet and books and journal articles that provide techniques and advice about how to explore these various aspects of commercialization.

In addition to determining how to get a product to market, once the product is there, one next needs to understand how to operate a business, hire personnel, figure out where to place the business (i.e., rent office space), and then how to handle the business finances. This entire spectrum of responsibilities moves one into the business environment, where many resources, instructional offerings, and consultants are available for guidance.

U OF U OFFERINGS—STAGE 4

As a U of U affiliate, one receives assistance on starting a new business from the TVC as well as the David Eccles School of Business. Numerous experts are available locally to address questions and provide processes for establishing start-up companies. There are several accelerator programs as well as entrepreneurial ones to become a part of if interested. These entities also assist with finding funding for new businesses, especially small ones, and with connecting a new business owner to local mentors and industry representatives. They can also help with identifying grant funding offered by the federal Small Business Innovation Research and Small Business Technology Transfer agencies. Assistance with identifying these funding agencies, writing grants, and administrating the funding is provided by the U of U Office of Sponsored Programs and the Grant Writing Service located in the Marriott Library. The EHSL and Marriott Library offer numerous resources on grant writing and funding sources as well, including private foundations. The EHSL has hired an innovation librarian to partner with innovators within the health sciences colleges and schools (chapter 9).

SUMMARY

From this chapter, one can easily see that having a great idea is just the first step toward becoming a successful innovator and entrepreneur. Knowing how to develop one's idea; strategize its value and the solution it provides to a problem; identify who will use the product; determine how to create and manufacture it, produce it, and sell it through distribution channels; and then manage the revenue generated and the expenditures associated with a product is a very complex cycle that needs to be understood and addressed early on. Luckily, most innovators are not alone and can avail themselves of numerous resources and experts to help guide them through these innovation stages.

This chapter has provided the general framework that needs to be followed. The remaining chapters in this book go into greater detail about many aspects that were just highlighted here. Remember that a good idea is key, but a great strategy for implementing the idea and developing it is needed in order for the idea to be *great*. Many good ideas die on the vine due to lack of follow-through or knowledge about how to take the idea to market. Use the multitude of resources and expertise that are available to you to become a successful innovator!

NOTES

1. Alexander Osterwalder and Yves Pigneur, *Business Model Generation: A Handbook for Visionaries, Game Changers, and Challengers* (Hoboken, NJ: Wiley, 2010).
2. Libraries Innovation Team, University of Utah, "Innovation: Welcome," http://campusguides.lib.utah.edu/innovate (accessed October 16, 2016).
3. Libraries Innovation Team, University of Utah, "Innovation: Business Model Canvas," http://campusguides.lib.utah.edu/innovate/bmc (accessed October 16, 2016).

4. Tallie Casucci, Erin N. Wimmer, and Jean P. Shipman, "Business Model Canvas Meets Evidence: The Intersection of Innovation Tools" (paper presented at the VentureWell Open 2016 Conference, Portland, OR, March 4, 2016).

5. Jean P. Shipman, Roger A. Altizer, José Zagal, Tallie Casucci, and Tina Kalinger, "Librarians as Matchmakers: Using Dating Sites as a Model for Collaboration" (paper presented at the annual meeting of the Medical Library Association, Austin, TX, May 18, 2015).

6. J. Willard Marriott Library, University of Utah, "Patents, Trademarks and Copyright: Home," http://campusguides.lib.utah.edu/patents (accessed October 16, 2016).

7. J. Willard Marriott Library, University of Utah, "Standards (ASTM, Codes, IEEE, Industry and Technical): Home," http://campusguides.lib.utah.edu/c.php?g=160326&p=1050391 (accessed October 16, 2016).

Chapter Three

Synapse

A Place Where Ideas Collide and Collaborations Congeal

Jean P. Shipman

OVERVIEW

The University of Utah (U of U) installed its Center for Medical Innovation (CMI), including a digital therapeutic games and app laboratory, a fabrication laboratory, and a skills center, within the Spencer S. Eccles Health Sciences Library (EHSL) to create a unique collaboration space for innovators, entrepreneurs, and industry to meet, share ideas, and produce medical devices, games, and apps. This co-location of information and innovation has proven valuable, as it gives all involved rapid access to the information needed to create context for games and apps and gives device generators the tools and facilities needed to produce prototypes of their ideas. How did this collaborative space, the Synapse, come to be? This chapter outlines the design process, benefits, lessons learned, and best practices gleaned for others who wish to follow suit.

BACKGROUND

In the summer of 2012, the U of U officially launched the CMI. This center was created by the president of the U of U to "fuel . . . collaborative effort between the University of Utah Health Sciences Center, The David Eccles School of Business, The College of Engineering and The Technology Venture Development Program."[1] The mission of the CMI is to provide "formal education programs, faculty and student project development, and support and facilitation of device development and commercialization. The Center creates a one-stop-shop environment that assists both the novice and experienced innovator through ideation, concept generation, intellectual property, market analysis, prototyping and testing, business plan development, and commercialization."[2] The CMI sponsors several educational events in specific topics of interest to faculty and students. It offers seed grant funding for faculty and students through several programs, including I-Corps and Bench to Bedside (B-2-B).

The CMI needed a home, and in July 2012, the CMI executive director asked the EHSL director if she could consider having it housed in the EHSL. The EHSL director was thrilled with this idea, as the EHSL had already established itself as an innovation and discovery center and had *Innovate* as one of its three tag words. The CMI director and the EHSL director met to establish ideas for how to embed the CMI within the EHSL.

At its origin, the CMI had an office for its new program manager and hall space right outside of the office on which to brand its name. It also repurposed three student study rooms into group meeting rooms for use by the CMI competition students and others wishing to work in groups on device creation.

THE GAPP LAB

In late 2013, the CMI executive director inquired whether a hands-on training laboratory used by the National Library of Medicine Training Center of the National Network of Libraries of Medicine could be repurposed to create a gaming laboratory for graduate students. This was accomplished, and in January 2014, The GApp Lab (Therapeutic Games and Apps Laboratory) was formed with 20 graduate students from the U of U's Entertainment, Arts, and Engineering Program (see chapters 8 and 12). These students were given fellowship stipends to create digital therapeutic games and apps.

The training lab was transformed overnight into The GApp Lab and has been running consistently ever since. The students utilized the existing instruction tables and chairs but had to acquire their own computers to provide more memory and graphics capabilities than what were supplied by the training lab computers. They repurposed a library office to create their own lounge, furnished with Ikea furniture. They also bought one-foot cubes to create a temporary storage wall to house their supply of food and beverages to fuel the students through long hours of design.

SYNAPSE CONCEPTION

Concurrently in 2012, the EHSL and the CMI executive director had been working with the Lean program initiated by the senior vice president for health sciences to encourage U of U Health Sciences faculty to become more efficient with processes and to reduce unnecessary waste. U of U Health Care had interest in implementing Lean processes throughout the hospital and clinics and were eager to create a space where Lean teams could *deep dive* into topics to derive solutions that would save on costs and improve services, quality, and ultimately value.

The EHSL director had mentioned to the CMI executive director that she could foresee the day when the compact shelving stacks of books and journals that were stored in the basement level of the EHSL would be removed to create more space. Months later, the CMI executive director inquired whether that time could be immediate, as the hospital would be willing to give funds to help repurpose the space. Thus became the conception of the Synapse, although the idea for what would be housed within it would evolve over time.

Another innovative group at the U of U, Spark (http://spark.utah.edu/spark-labs), had also inquired about being located within the EHSL at the same time that the CMI executive director had approached the EHSL director. The EHSL director asked the Spark director for help designing the new deep-dive facility, as he was also an architect. Several designs resulted, one of which included retaining some of the movable compact shelves to serve as walls to create group meeting spaces of varying size based on group numbers. This design was also desirable, as the funding to remove the compact shelves in their entirety was not available. As all of these deliberations required time, the senior vice president of health sciences had meanwhile decided not to create a deep-dive space in the EHSL due to Lean having become part of the daily work of the Health Care system, and thus separate meeting spaces were no longer needed for teams to convene.

The CMI executive director and the EHSL director met again and decided to remove the stacks completely to be able to create space for innovation students, faculty, and industry mentors to collaborate instead. The EHSL director attempted to locate remote storage for the materials in the U of U's academic library, in a robotic storage facility, and also in a data center located downtown, neither of which came to fruition. The EHSL director was able to acquire additional funding from the hospital, for a grand total of $370,000, to remove the stacks and hire an architect to redesign the space for a remodel. The EHSL contributed additional library endowment revenue to acquire digital back files for the most highly used journals stored in the space and to pay for the actual remodel and new furnishings for the space that was named the Synapse (total cost was $1,096,020).

Thus began the laborious project of the EHSL staff to identify high-use items housed in the compact stacks, acquire their digital back files, and find homes in other libraries for some of the materials, including the National Library of Medicine. An article within *Collection Management*[3] outlines the process undertaken by the EHSL staff to remove the collection within a rapid nine-month period of time.

With the print materials out of the way, the stacks were able to be removed, a new space design was created, and all was ready to roll for a remodel, except a donor appeared who wanted to create a surgical simulation center for the U of U. The CMI executive director asked the EHSL director to consider how to incorporate such a center within the EHSL. The architect went back to the drawing board, and another remodel plan surfaced. This time, the EHSL director contributed significant input based on recent visits to other team spaces and to furniture company showcase facilities. She also requested that the space be created with flexibility in mind, as the space had already gone through numerous designs based on changing needs. She provided ideas that were incorporated in the final design (see figure 3.1). This final design was not final, however, as the simulation center donor decided not to provide his funding after all, and a new chair of surgery came on board and questioned the idea of placing a simulation center within the EHSL. The simulation center space was then repurposed to accommodate The GApp Lab students with electrical power being dropped from the ceiling to facilitate four pods of multiple workstations, and storage closets, cupboards, and an office were removed from the original simulation center plans.

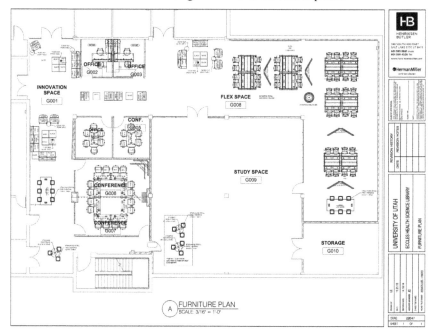

Figure 3.1. Final Synapse Design. *Henriksen Butler.*

CONSTRUCTION AND DESIGN CHALLENGES

With the true final plans drawn, construction of the Synapse began in early 2014. One of the very first issues that surfaced was that the walls were not designed to go to the ceiling deck, as DIRTT walls (http://www.DIRTT.net), an easily movable wall that hangs from ceiling tile grids, had been planned when the deep-dive concept was alive. Noise and sharing of team information was not an issue when the space was being designed for Lean teams, but with innovators and those wishing to acquire patents on their idealized products, team collaboration rooms needed to be more soundproof. The architect and the U of U physical project manager decided to replace the DIRTT walls with double-paned glass walls that could be lined with a special coating to provide some privacy as well as enable writing on both sides of the glass walls. These new walls reached to the ceiling deck and provided substantially more sound protection. However, they increased the cost of the remodel and reduced the flexibility of being able to move the walls when desired. Tradeoffs would become the name of the game as construction continued.

The next design change that resulted as construction proceeded was with a café area that was being built into the south wall for individuals to use during the day for lunches and breaks and for receptions and events during the evenings. Plumbing for a sink for this café area would have been costly, as no water existed in that space. A relocation of the café area was made to place the sink closer to the floor's restrooms and thus water sources. This relocation resulted in decreasing the cost of the remodel because piping did not need to be run to the south wall.

The resulting café area actually solved another problem in that it now occupies an empty foyer area that had previously been underutilized and is more centrally located. Students find it to be an attractive break space, and all entities of the Synapse use it for their kitchen and food supply area (see figure 3.2).

A third major remodel design change that resulted in a design flaw was with the simulation center. While at the time of the remodel this space was not going to be a simulation center, it resulted in becoming the Skills Center. The door that had been installed on the east wall to permit The GApp Lab students egress into the bay areas needs to remain unlocked for emergency exit purposes for individuals using the back bays. This is an issue since the room contains expensive simulation equipment and is frequently not manned. Several remedies are being considered, none of which is inexpensive. Had the final purpose for the space been decided prior to construction, this need for emergency egress would have been taken into account with the remodel.

Figure 3.2. Synapse Café.

SYNAPSE OPENING

In June 2014, the Synapse officially opened. During the remainder of 2014, additional small items were added, including an attractive clock, a Synapse sign and directories, bay instruction signs with bay numbers and how to acquire video-monitor connection cables and whiteboard markers, and a room scheduling video monitor that offers users of the Synapse a guide to meeting rooms.

The final space is highly functional and very inviting and bright for being in the basement of the EHSL, now fondly dubbed the *Garden Level*. It includes three offices, one small glass-walled conference room, and a large glass-walled conference room that can be divided with a glass soundproof partition to form two smaller conference rooms that can be further subdivided with mobile telescopic whiteboards to create four small meeting rooms. The CMI program director moved into one of the offices, an innovation librarian and the EHSL director moved into the others. The large Skills Center space is used by different departments within the School of Medicine, primarily Surgery. While storage space was not built into this center, rolling storage cabinets were acquired and are rented to departments. Six group bays exist within an L-shaped space outside of the Skills Center. These bays are divided by mobile telescopic whiteboards that expand from two to six panels.

Each bay has four to six rolling tables that combine into one with movable lightweight chairs and a wall-mounted video monitor. The video monitors can be individually used or chained to transform the L-shaped bay space into a lecture hall where the chairs and tables can be arranged theater or classroom style. Bays are used on a first-come, first-served basis, whereas the conference rooms can be reserved in advance. In addition to the areas described, there are three lounge areas and several casual two-person seating areas for ad hoc conversations or quick meetings. A donor wall with red and white tiles creates a unique design element and is often used to provide an interesting backdrop to photos and videos. Part of the south wall is mirrored to provide the illusion of extra space, and two stand-up electric-powered tables with 12 stools provide laptop users temporary work space. A fun item within this workspace is a *wobble* chair that spins to circulate the blood and clear the head so that new ideas can transpire, and the café described earlier exists for all to enjoy!

ADDITIONAL SYNAPSE ENTITIES

Gary L. Crocker Innovation and Design Laboratory

In February 2015, a fabrication laboratory called the Gary L. Crocker Innovation and Design Laboratory opened for business (chapter 6). Within what was the prior EHSL mailroom, the lab now contains 3D printers, body scanners, band saws, CNN mills, and lathes as well as electrical soldering equipment for students, faculty, and others to create prototypes of their medical device designs. Ideas literally solidify within this lab with the help of engineering staff and student fellows. The lab is open during the week with regular hours and after hours on demand. As the competitions deadlines draw near, the lab hums and products rapidly emerge, enabling designers to immediately alter their products or to showcase them during the competitions. For more information about the equipment offered by the lab and its hours and services, refer to http://library.med.utah.edu/synapse/fablab.

Skills Center

The challenging space that led to many redesigns of the Synapse did indeed turn into a simulation or skills center shared by many departments of the School of Medicine, including the lead department, Surgery. While still being outfitted, the Skills Center includes a da Vinci robotic surgery training simulator, several task trainers, and numerous simulators of varying complexity. The Skills Center also hosts classes to train residents and medical students as well as other professional community members. The proximity of the bay areas is appreciated, as training often exceeds the 2,000-square-foot space allocated to the Skills Center itself. Students can clump into groups within the bays to view videos, slide decks, and other instructional visuals. They can also reserve nearby conferences rooms for group deliberations for large meetings. Again, the Synapse space is proving to be flexible enough to meet the demands of another Synapse entity.

POST-REMODEL CHANGES

In February 2015, the EHSL director moved her office into one of the smaller conference rooms to be better able to see how the Synapse space was being used and to identify the remaining needs of innovators and occupants. Being located within the Synapse has offered many benefits, including being able to easily converse with CMI staff and students and to be readily available to address facility needs and to ensure that the space is available for both EHSL and CMI users. There is nothing like being embedded to be able to see how a space is being used and what changes are needed to accommodate users' preferences.

In October 2015, the CMI expanded into what had been the space occupied by the Center for Clinical and Translational Science. This additional space provides a large office for the CMI executive director, an office for The GApp Lab director, a shared office for the director of engineering who runs the fabrication laboratory and several student fellows who assist in the lab, and an office for a development officer, in addition to another large conference room. This area adjoins to the Café and The GApp Lab spaces. An additional office is shared by others within The GApp Lab. The entire Garden Level of the EHSL now offers innovators and EHSL users a joint respite and gathering place to share ideas, get assistance, develop professionally, and meet with others to create new and exciting products, games, and apps. In fact, it has been designated a *safe space* for the Health Sciences campus.

SUMMARY

All in all, the Synapse has been optimal, and despite the three main design changes during its construction process, not many adjustments are needed. The space provides flexibility to accommodate small or large groups, and the glass walls and mobile whiteboards provide opportunities for users to capture ideas as they flow and to share their ideas with others.

What does the resulting Synapse space look like? Figures 3.3 through 3.5 show some photos of the Synapse. A video of the Synapse space may be found at https://youtu.be/vQ3pEHyhqU0. See how the EHSL basement has been transformed into a bright, collaborative collision space for many to enjoy and be productive.

Figure 3.3. Synapse.

Figure 3.4. Synapse Donor Wall and Lounge Area.

Figure 3.5. Synapse Group Bay Area.

NOTES

1. Center for Medical Innovation, University of Utah Health Sciences, "What We Do," http://cmi.uofuhealthsciences.org/about_us (accessed October 19, 2016).
2. Ibid.
3. Christy Jarvis, Joan M. Gregory, and Jean P. Shipman, "Books to Bytes at the Speed of Light: A Rapid Health Sciences Collection Transformation," *Collection Management* 39, no. 2–3 (2014): 60–76, doi:10.1080/01462679.2014.910150.

Chapter Four

Making the Makerspace

The Nexus Collaborative Learning Lab

Jennifer Herron and Kellie Kaneshiro

In 2015, the Ruth Lilly Medical Library, Indiana University, designed a makerspace including a 3D printing service. The library wanted to ensure the success of the planned space and contacted other university departments for feedback and to foster collaborative partnerships in the process. The goal of the makerspace was to promote collaboration not only between medical students but also between other health professions and schools within the university. The makerspace officially opened in March 2016, and it is fittingly called the Nexus Collaborative Learning Lab.

BACKGROUND

The Indiana University School of Medicine is the largest medical school in the United States (http://medicine.iu.edu/education/iusm-campuses). The incoming 2016 class had more than 1,400 medical students and almost 400 students in the PhD, MSMS certificate, and master's programs.[1] Students are spread throughout nine campuses across the state of Indiana: Evansville, Bloomington, Indianapolis, Terre Haute, Muncie, West Lafayette, Fort Wayne, South Bend, and Gary. The Ruth Lilly Medical Library is located on the Indianapolis campus, and it serves all of the other campuses through virtual reference service, e-resources, and scheduled on-site visits by librarian liaisons.

Research is a top priority of the Medical School, with more than 800,000 dedicated square feet of research space. In 2015, more than $300 million in research grants was awarded to faculty and staff at the school.[2] Indiana

University as a whole is also committed to research and developed the Grand Challenges Program in 2015 with the aim of addressing "major and large-scale problems . . . that can only be addressed by multidisciplinary teams of the best researchers."[3] The school submitted a Precision Health Initiative proposal and is the first recipient of a Grand Challenges Program award. The Precision Health Initiative's goal is to "transform biomedical research, health care innovations, and the delivery of health interventions in Indiana."[4]

The Medical School is also part of a statewide collaborative called the Indiana Clinical and Translational Sciences Institute (CTSI), which "facilitates the translation of scientific discoveries in the lab into clinical trials and new patient treatments in Indiana and beyond."[5] The CTSI is a statewide collaboration of Indiana University, Purdue University, the University of Notre Dame, and a mix of public and private partnerships. The library's makerspace supports collaborative learning not only for the Grand Challenges Program and the Precision Health Initiative but also for interprofessional engagement that is listed as a goal of the school's new curriculum.

The decision to build a makerspace resulted from the library Technology Team's massive action plan that identified areas and services for enhancement. With the Medical School working on a curriculum renewal, the library wanted to update and enhance its services in order to address the ever-changing medical field. Incorporating 3D printing along with the concept of collaborative learning into the makerspace was a natural fit.

The team explored how to start a 3D printing service by researching 3D printing and reaching out to known experts in the field. A review of other libraries with similar spaces and services ensued. A librarian with 3D printing knowledge recommended Cube printers due to their ease of use and personal success with their operation. A library makerspaces listserv was also consulted to obtain additional feedback. Ultimately, the library purchased two Cube printers: a CubePro and a Cube 3.

The need for 3D printing in medicine and medical education was investigated in order to best identify how to promote the service and maximize use of it in the makerspace. A literature search on 3D printing provided useful information on how 3D printing could support medical education. Attendees of the 2014 Surgical Education Week Conference responded to questions as to whether 3D-printed models would assist in teaching abnormal pathology; 90 percent of respondents agreed, and another 90 percent agreed when asked similarly if the models would be useful in teaching basic anatomy.[6] In regard to medical education, all respondents agreed that medical schools could benefit from using such models in their curricula. Furthermore, Jones and colleagues noted, "By creating models of actual pathologic disease processes, medical students and junior residents might gain a better understanding of normal anatomy and pathologic disease."[7]

Researchers at Monash University's Clayton Campus found similar evidence for the use of 3D-printed models in medical education. Fifty-two first-year medical students took a pretest to determine baseline cardiac anatomy knowledge and were then randomized into three groups. One group used cadaveric models, another used 3D prints, and the third group used a combination of both. Each group received a task sheet with learning objectives to complete and, on completion, were given a posttest. The posttest scores were significantly higher ($p = 0.010$, adjusted $p = 0.012$) in the group using the 3D models (mean of 60.83 percent) compared to the group using cadaveric materials (44.81 percent) or the group that used a combination of cadaveric and 3D models (44.62 percent).[8] Interestingly, researchers suspected students' hesitation with handling the cadaveric materials as potential explanations for the increased performance in the 3D printing group. Furthermore, researchers noted that the multicolored models also appeared to aid in students' learning comprehension.

FIELD TRIPS

Members of the Technology Team attended the Medical Library Association's 2015 annual conference and noted that librarians presenting on 3D printing found that the biggest barrier to engaging students and faculty with 3D printing was attracting interest and breaking down the perceived difficulty of 3D printing. Understanding the needs and issues with starting a 3D printing service helped the team to better understand how to market this service, how the makerspace could support 3D printing, and how 3D printing could enhance the collaborative makerspace experience.

Touring 3D printing labs on campus and visiting other spaces with advanced learning technology generated ideas for what kind of technology and features the library wanted in the new space and helped with designing the makerspace. After visiting 3D printing labs across the university, the team identified the need for postprocessing tools to clean 3D prints. A visit to a learning lab at another library on campus also highlighted the need for flexible space that could easily be adapted to fit the needs of multiple types of events or activities, such as workshops, group projects, or independent study. Additionally, the team visited classrooms that were part of the university's Mosaic Learning Initiative (https://uits.iu.edu/mosaic). The Mosaic Learning Initiative began in 2015, and it promotes the use of technology in the classroom. Several *active learning classrooms* exist on the university's Bloomington and Indianapolis campuses. These *Mosaic Classrooms* offer students classrooms equipped with technology that facilitates engagement between the instructors and students.

Finally, a trip to the Advanced Visualization Lab (AVL) (https://rt.uits.iu.edu/visualization/avl/systems/iq-wall.php), operated through the Research Technologies Unit of University Information Technology Services, highlighted several technologies that the team decided to place in the makerspace. These include an IQ Wall, 3D scanners, virtual reality, and augmented-reality applications. The IQ Wall is a creation of the AVL's distributed visualization initiative and was custom installed using commercial hardware and open-source software for operation.

Since anatomy education was one target area needing support, specialized *virtual dissector* software was considered. To evaluate the software, faculty members from the Department of Anatomy were consulted and invited to attend virtual demonstrations of two software applications. The library coordinated and hosted the sessions to ask anatomy faculty for their input on the makerspace technology, including 3D printing.

Once a plan was drafted and approved, librarians worked with representatives from the Medical School's Space Planning and Organization Unit who partnered with the University Architect's Office on renovating the space and selecting the furniture.

COMPLETED SPACE

The Nexus officially opened in March 2016, with the last piece of hardware installed at the end of August 2016. The makerspace includes a variety of technology and features that enable collaborative work, independent study, and the hosting of library technology demonstration events.

The focal point of the Nexus is the 16-foot IQ Wall, the last piece to be installed. The IQ Wall is comprised of eight video monitors with a touch overlay. Once users log in to the IQ Wall, the overlay allows users to easily use the IQ Wall and facilitates interaction with the displayed content. The connected screens form one large monitor using a software program called *Mosaic* to span the screen. With such a large screen, multiple windows and programs can be opened and expanded, enabling PowerPoint presentations to display simultaneously with supporting presentation materials. Laptops and other devices can also connect to the IQ Wall, and the screen can be split and divided to show content from multiple devices, including the main computer that operates the IQ Wall. Programs such as *Paint* also enable multiple users to interact simultaneously, allowing more collaboration.

The Nexus includes two mobile media:scapes that allow devices to connect and display content through VGA and HDMI connections. Adapters were purchased that allow mobile devices, such as tablets and smartphones, to be connected and displayed through the media:scape's main monitor. Up to four devices can be connected to each media:scape, with only one device screen being displayed at a time. Users can quickly switch between projected screens by pressing the *puck* connected to their device in order to take control of the monitor.

The media:scapes are mobile and can be moved about the space to accommodate groups of any size. This flexible and mobile trend continues with the furniture within the space. The mobility flexibility of the space was one of the most desired features (see figure 4.1). Tables and chairs can similarly be moved anywhere throughout the room. This enables students to define their own spaces for independent study while also allowing group meetings, workshops, and other events to occur within the Nexus. The movable furniture and mobile media:scapes have enabled librarians to hold workshops and demonstrate different library resources while still allowing nonparticipating students to work on projects or continue to study.

The 3D printing lab is located inside the Nexus and includes three printers—a CubePro, a Cube 3, and a Makerbot Replicator, gifted by University IT. Tools to enable original 3D products are also available. Two scanners support discovery and creation via 3D printing. A Cube Sense and iSense scanner circulate so that researchers can find new ways to view and manipulate their research.

An Indianapolis campus–based Medical Student Technology Committee recently funded medical students' request for GoPro cameras. Currently, three GoPro cameras can be borrowed. These cameras are maintained by Nexus student workers, with anticipation that some of the multimedia pro-

Figure 4.1. Nexus Mobile Furniture.

jects produced using the GoPro cameras will be displayed on the IQ Wall or the media:scapes.

The library's makerspace includes a small adjacent breakout room for groups to meet privately and explore the technology in greater detail. Currently, this room is used to showcase and explore the *BodyViz* software. This software is installed on the IQ Wall in the main Nexus and in the breakout room on a 98-inch touch-screen monitor.

Additional technology to be added to the Nexus includes head-mounted displays for virtual and augmented reality. The space already includes examples of augmented reality by providing targets for the *Anatomy 4D* app and other available health-related augmented-reality apps. Currently on order are Virtuali-Tees, which allow students to learn more about anatomy while having fun.

SERVICES OFFERED

As the IQ Wall was only recently installed, the library has not yet had the opportunity to implement the majority of services and features it has planned for the Nexus space. It will offer workshops, drop-in sessions, and *Tech Talks* soon. Meanwhile, students use the space to study, work on group projects, and review course materials.

A 3D printing workshop is conducted in the Nexus. A very basic introductory session on 3D printing and how the printers work is offered, as are tours of the 3D printing lab. Additionally, 3D-printed models are displayed inside the Nexus for students to interact with so that they learn what the 3D printers can produce.

The media:scapes are used to demo different mobile applications. Connecting mobile devices to the media:scapes allows for the devices' screens to be displayed so that large groups can see the apps. During the Nexus open house, an iPad connected to a media:scape enabled visitors to explore features and test various medical and health apps. Students have also connected their devices to the media:scapes to review lectures and collaborate with classmates on projects.

When designing the makerspace, the library wanted to include software to support anatomy education. Different *virtual dissectors* were compared with *BodyViz* being selected for installation on the IQ Wall. *BodyViz* is a program that comes preloaded with scans that generate 3D models and also enables DICOM data to be uploaded. *BodyViz* features include the ability to dissect the model—slicing through layers to view specific sections. This program uses an Xbox controller that quickly catches students' eyes when this software is presented to groups.

Along with supporting the learning needs of students, the Nexus also serves their wellness needs. One resource that students have exhibited interest in is the *Calm* app and website. Loading the website https://www.calm.com onto the IQ Wall will bring up a meditative scene accompanied by music. This website is enjoyed by Nexus workers. As a result, the library is working with the AVL to include this website as an IQ Wall screen saver when not in use (see figure 4.2).

Furthermore, the library is interested in collaborating with the Medical Student Education lead advisers (conveniently located next door to the space) on additional student wellness events. So far, there have been two events that utilized the IQ Wall: an Electronic Residency Application Service (ERAS) Twitterchat and a session where a lead adviser guided students through the ERAS application process.

FUTURE PLANS

Many ideas for events and services have yet to be acted on for the Nexus. The library wants to better connect students across the many campuses it serves so that they may participate in library events. One planned event is a research poster session where all students can submit posters for Wall display. In the same light, the library also would like to feature 3D scans of some items from its History of Medicine (HOM) collection and create a virtual museum type of event to highlight the HOM items. Additionally, the library is reaching out to technology-focused medical student interest groups

Figure 4.2. Nexus IQ Wall.

to offer tours, demos, and the Nexus as a potential meeting space. Art events and other stress-busting activities are planned for the future.

The Nexus serves as the meeting location for a new, interdisciplinary 3D printing interest group formed in early 2016. The group's size has doubled since its first meeting in February, growing from 10 members to more than 20 from 11 different departments. The group consists of members who initially met during the library's tours of different 3D printing spaces throughout the university. The group plans to hold recruitment events to establish different student interest groups. Additional future plans include hosting a University Makerfaire in the spring of 2017.

Student workers staff the Nexus. These students have a background in 3D modeling and are enrolled in a technology-focused program. Staffing the Nexus, even part-time, provides additional technical support to visitors, providing someone to demonstrate the technology immediately as opposed to having to arrange a consulting time with a trained librarian. Furthermore, these students are able to provide their creative input on additional uses for the technology.

The Data Management Librarian helps to support and promote use of the Nexus and its services. The IQ Wall is a prime resource for data visualization. The library is identifying a niche in research on campus with its 3D printing service as a means for researchers to interact with and manipulate their data. Programs such as *Tableau*, *ParaView*, *VisIt*, and *VMD* provide researchers this opportunity to display their data and create visual renderings to give insights into their data.[9]

Showcasing faculty and student work and hosting simulated TED talk events in the space are future ideas for the Nexus. Makerhealth is a new breed of makerspaces in the medical and health care field. Could the Nexus become a makerhealth space to encourage student and faculty involvement with makerhealth events and activities? Makerhealth organizations encourage student and professional innovation and are focused toward medical innovation across the professions. VentureWell is one organization that helps to promote medical innovation through its biomedical engineering competition called the *BMEidea* (https://venturewell.org/whatwedo). For more than 10 years, the *BMEidea* competition allows entries from undergraduate and graduate students from a variety of different backgrounds and encourages interdisciplinary teams to submit devices, products, or other technology in such areas as surgery, home health care, diagnostics, therapeutics, and preventive medicine.

Other similar organizations exist that spur medical innovation by providing tools and spaces for medical professionals to gain the skills needed to design and prototype their ideas and put their gained skills to use. Makerhealth is an organization that connects people inventing, or making, in the health field and provides access to their website to explore projects and provide training resources to learn how to complete personal projects (http://www.makerhealth.co).

CONCLUSION

Planning and construction of the Nexus took 14 months, and expansion of its capabilities continues today as the library investigates new technologies and services for the space. The name Nexus was selected to emphasize the space's intention—to connect or link together. Students have used this space for collaborative study for exams and projects. The library is excited to offer such a space that encourages collaborative learning, and it hopes that the Nexus will continue to foster such connections in the future.

NOTES

1. Indiana University, "Indiana University School of Medicine: Fact Sheet for 2015–2016," http://medicine.iu.edu/files/2014/5495/4805/IUSM_Fact_Sheet_2015-16.pdf (accessed October 16, 2016).
2. Ibid.
3. Fred H. Cate, "Grand Challenges: About," http://grandchallenges.iu.edu/about.html (accessed September 12, 2016).
4. Indiana University, "Indiana University School of Medicine's Precision Health Initiative," https://medicine.iu.edu/research/areas-of-expertise/precision-health-initiative (accessed September 12, 2016).
5. Indiana Clinical and Translational Sciences Institute, "About Us," https://www.indianactsi.org/about/aboutus (accessed August 15, 2016).
6. Daniel B. Jones, Robert Sung, Crispin Weinberg, Theodore Korelitz, and Robert Andrews, "Three-Dimensional Modeling May Improve Surgical Education and Clinical Practice," *Surgical Innovation* 23, no. 2 (2016): 189–95, doi:10.1177/1553350615607641.
7. Ibid., 193.
8. Kah Heng Alexander Lim, Zhou Yaw Loo, Stephen J. Goldie, Justin W. Adams, and Paul G. McMenamin, "Use of 3D Printed Models in Medical Education: A Randomized Control Trial Comparing 3D Prints versus Cadaveric Materials for Learning External Cardiac Anatomy," *Anatomical Sciences Education* 9, no. 3 (2015): 213–21, http://onlinelibrary.wiley.com/doi/10.1002/ase.1573/full (accessed October 22, 2016).
9. Brian Suda, "The 38 Best Tools for Data Visualization," *CB—Creative Bloq: Art and Design Inspiration*, http://www.creativebloq.com/design-tools/data-visualization-712402 (accessed October 14, 2016).

Chapter Five

Supporting Institutional Strategic Directions and User Needs through Library Collaborative Spaces

Mary Joan Tooey

LIBRARY AS COLLABORATIVE SPACE

When the Health Sciences and Human Services Library (HS/HSL) at the University of Maryland, Baltimore, opened in 1998, it was one of the largest health sciences libraries in the United States. At that time, printed books and journals were still the norm. The building was designed for 15 years of growth space in order to reach 85 percent capacity. That capacity was never needed, as the evolution and revolution of e-journals and e-books was just around the corner.[1]

The HS/HSL was also designed as a teaching and collaborative space with three classrooms, four conference and boardrooms, 45 small-group study rooms, and more than 900 seats within easy access to power and hardwired data connections. While the need for stack space diminished, the need for collaborative workspaces did not.

This has proven advantageous, as the basic principles inherent in the design focused on a people-centered building, inclusive of a coffee shop, an art gallery, and an adjoining campus center. An advanced technology infrastructure has supported multiple renovations and iterations of the basic user-centered design. With plenty of light and color, people were placed near the windows, and collections were protected by their central location in the building. As stacks were no longer needed for expanding collections, large swaths of shelving were removed. The open areas were repurposed for additional office spaces and new types of collaborative spaces and programming.[2]

From the time of the building's opening, the HS/HSL has been a focal point for the university. It is a welcoming, neutral space away from the silos of the schools, programs, and units and accessible to the entire university community. It has served as a hub for student collaboration and activity. While the building is enjoyed by the entire university community, students are the primary users of the building. Annual surveys of the students are undertaken to determine their needs within the building. Whether it is rolling whiteboards or new water fountains configured to easily fill water bottles, students drive many of the spatial changes. These surveys also guide renovations, services, and, in many cases, new collaborative initiatives. The HS/HSL's strong connections to schools and programs through its liaison program affirm student and faculty needs. Through the liaisons, the HS/HSL team is aware of existing services in order to respond to needs and avoid duplication of programs with schools and other units and unnecessary expenditures of staff and financial resources. The executive director also serves as the associate vice president for academic affairs. Through her position, she is knowledgeable about services available within the schools or other university units and is frequently part of university-wide discussions regarding collaboration, collaborative spaces, and academic program needs.

THE PRESENTATION PRACTICE STUDIO

One request from student surveys indicated a need for a space where students could practice and refine presentations for classes, whether by themselves or for group projects. Faculty had also indicated an interest in a place where they could record, film, and edit online classes or presentations. Out of those requests, the Presentation Practice Studio was born (figure 5.1). As an institution that celebrates its diverse campus composition and collaborative work, creation of this space made a tremendous amount of sense.

These types of requests present learning opportunities for HS/HSL faculty and staff as the HS/HSL team engages in studying an idea thoroughly and then works on the project concept from inception to completion. These projects support the development of new skills and expertise and advance the library's mission.

In the case of the Presentation Practice Studio, there were a number of challenges. What was the current state of the art for studios of this type? Where should it be located? From where would the funding come? What staff would support the facility? What services would be offered? How much would be self-service with minimal intervention? What policies and procedures would ensure its success?

The standard protocol at the HS/HSL when beginning new projects like this is the development of a white paper exploring those issues and teasing

Figure 5.1. Presentation Practice Studio.

out any other considerations. A project team is then assembled. In this case, there was representation from the Services, Resources, Regional Medical Library, and Technology Divisions. The Project Team, formed in June 2008, focused more broadly than just on a Presentation Practice Studio in order to inform future development of collaborative spaces. With the completed report, the team identified seven functions and services including the recommendation for the studio.

Recommendations included flexible work areas on the first and second floors, a multimedia design studio, scalable videoconferencing capabilities, upgraded study rooms, virtual study spaces on the library's website, and expansion of services to include equipment loans and workshops on collaboration tools. The report commented on the availability and neutrality of the library as a collaborative learning space and identified factors such as lighting, security, and soundproofing and the importance of embracing user-centered design to ensure an inviting, user-friendly space.

Along with the Presentation Practice Studio, several of the other recommendations, such as videoconferencing and upgraded study rooms, were implemented. Others, such as the multimedia design studio, were put on hold for financial and spatial reasons. Many of the functions of a multimedia design space were made available within the Presentation Practice Studio.

Still others have not been implemented due to changes in technology or need.[3]

The recommendation for the Presentation Practice Studio stated, "Design a studio for students, faculty, and staff to practice presentations. Equip with a flat screen monitor and podium; allow for video capture and playback."[4] Appendix B of the report expanded on the personnel and equipment needed, including future upgrades and ongoing maintenance requirements, and also identified the necessary policies and procedures to be developed.[5] However, the main questions of location and funding remained to be answered. A lesser-used, interior 300-square-foot study room was identified as a best choice for location for the studio. This study room contained a 77-square-foot closet that could be converted to a control and editing room. The renovations for the Presentation Practice Studio were funded by a donation from a family foundation that had previously funded HS/HSL educational technology projects. Interestingly, after the studio opened, one of the foundation executives needed to convert some 35-mm film to slides for a presentation and became a satisfied customer of his foundation's funded space.

Construction on the studio began in the fall of 2010 and was completed in January 2011. The room is soundproof and equipped with a 42-inch LCD panel, a podium on wheels, and a laptop computer. The control/editing room contains a computer with two monitors, cameras, lighting, tripods, a sound mixer, microphones, and stands. Professional and consumer video and audio editing software are offered.

During the construction period, an Implementation Team created a service/staffing plan, an online reservation system, instructional handouts for use of the studio equipment, and policies and procedures for use of the space. The team included HS/HSL information services and information technology staff. The information services staff's current duties include management of online reservations. As the first line of support, they offer basic assistance with using the video camera to capture and play back presentations. Information technology staff offer expert assistance, especially in the use of editing software. Users supply their own audio or video storage devices. The studio is available seven days a week during business hours by appointment only. Its website contains detailed information about the studio components and capabilities, the online reservation system, and a list of equipment and guidelines for use (http://cal.hshsl.umaryland.edu/booking/studio).

The studio officially opened for business with a party in March 2011. In the intervening five years, the studio has been well used for a variety of reasons. Student and faculty use it equally, mainly to practice presentations. Deans have practiced and taped state-of-the-school addresses. Faculty have prepared online instruction. Students have worked on promotional videos for their schools, programs, and end-of-the-year skits and follies. It has been used to support public speaking training for high school students in a grant program. Webcasts have been launched from the control room. During fiscal year 2016, the studio was booked 233 times with faculty representing the largest group of users. As of July 2016, 93 faculty, 70 students, 30 postdoctoral students, and 40 staff have used the space. The average length of usage is two hours.

The popularity of the studio, the aging of the equipment, and the increased use of technology in courses have created challenges. One is sustainability from both a personnel and a technology renewal standpoint. Staff have to be trained to support the studio and learn any new technology or equipment added. Staff turnover and loss of expertise create a need to ensure that appropriate staff with media experience are hired and trained thoroughly. Instructional technology specialists have now become key members of the HS/HSL team. Aging equipment must be replaced and renewed. Keeping up with the constant emergence of new technologies is ongoing. Finding funding to continue to support this initiative may be a future challenge.

The need to increase and support this type of space will only continue to grow. The university plans to expand its footprint in online education. Interprofessional education is firmly part of the university's ecology. A center focused on faculty development in the use of new technologies and pedagogy is planned. These and other initiatives will require additional attention to presentation/practice sites. Adding or expanding such spaces will require thoughtful examination of expansion space for additional studios and staffing to support them.

THE INNOVATION SPACE

Although the first patent for a stereolithography apparatus was issued to Charles Hull in 1986,[6] 3D printing did not become an affordable, small-scale production modality until 2009 with the first commercially available 3D printing kit. In 2012, 3D processes were introduced in an affordable format. Since that time, applicability of 3D printing in many sectors has exploded. It was only a matter of time before the health care industry paid attention. Steven Leckart's article posted August 6, 2013, on the *Popular Science* website titled "How 3D Printing Body Parts Will Revolutionize Medicine," was highly approachable and understandable, illustrating possibilities in health care utilization.[7]

Concurrently, the rise of the *maker* movement in the middle of the first decade of the 21st century encouraged entrepreneurs, inventors, hackers, and innovators to build and experiment on small scales, developing creative solutions to various problems. Soon makerspaces were springing up in communities, museums, universities—and libraries. The marriage of makerspaces and 3D printing seemed to be a match made in heaven and a need to be met at the university.

The university's 2011–2016 strategic plan strongly focused on innovation and entrepreneurship (http://www.umaryland.edu/about-umb/strategic-plan/2011-2016-strategic-plan). How could the HS/HSL support this focus? In a highly siloed environment, while 3D printing existed in some departments on campus, few people, particularly students, knew of or had access to these emerging technologies. It was incumbent on the HS/HSL as a space accessible to all to define what a public makerspace with 3D capabilities could mean for the university.

Using the tried-and-true methodology of assembling a task force, a library team began investigating the feasibility of locating a makerspace facility and services within the HS/HSL. The newly hired associate director for library applications and knowledge systems chaired the task force. Establishing an aggressive schedule, the task force began by looking at the state of the art in makerspaces in Baltimore and beyond, particularly in higher education institutions. There were field trips and site visit outreach to the maker community in the Baltimore area. The resulting *Makerspace Task Force Report*, published in July 2014, clearly defined and delineated the opportunities and challenges, recommending the creation of such a space within the HS/HSL.[8]

As with the Presentation Practice Studio, funding and space needed to be identified. Funding was again secured from an outside donor. Finding a location within the library was more challenging, as the vision for the space was for it to be publicly visible and easily accessible. A determination was made to make this an open-air innovation space rather than locating it in a separate room. This resolved many of the issues regarding air quality within an enclosed space. A locational opportunity presented itself when the Services Division decided to consolidate two service desks into one, freeing the former space for renovation. The very complicated transition process involved the reenvisioning of the Information Services Desk, decommissioning and removing the former desk; designing the makerspace; construction, ordering, and installing equipment; staffing and offering training; and working on the policies, procedures, website, and other publicity to support this new space. An advisory committee of outside advocates was formed. All of this needed to occur simultaneously and be highly coordinated.

Figure 5.2. Innovation Space. Health Sciences and Human Services Library, University of Maryland.

In April 2015, the library opened the first iteration of the makerspace, now named the *Innovation Space* or, as it is internally known, the *i-Space* (figure 5.2). More than 100 people attended the opening. Initially, the space included two 3D printers and scanners, molecular models, a button maker, and access to instructional software. Currently, the Innovation Space offers three 3D printers, two 2D scanners, a plotter for poster printing, access to more than 3,500 tutorials from Lynda.com, a DNA model, a molecule model, and a button maker.

By June 2015, primary responsibility for oversight of the Innovation Space was assigned to the new and emerging technologies librarian. Due to demand, interest, and a vision to make the space more interactive, by October the footprint for the Innovation Space was doubled. A more powerful 3D printer and 3D scanner were added along with more furniture and monitors in support of collaborative work within the space. The space was given a more industrial workshop look with pegboards, blackboard paint, hooks on which to hang tools and filament spools, and worktables and stools.

The new and emerging technologies librarian developed consultation services and a series of workshops introducing tools and programs to advance the expertise of users within the Innovation Space. In April 2016, an *Innovation Space Newsletter* was introduced. Subscriptions to this newsletter now exceed those of the *HS/HSL Connective Issues Newsletter*. An Innovation Space website aggregates all information about the space, including pricing, reservations, newsletter content, and announcements about programming and events (http://www.hshsl.umaryland.edu/services/ispace).

In fiscal year 2016, the first full year of operation, use of and interest in the Innovation Space grew. There were 61 unique clients and 90 total print jobs with production length ranging from one hour to almost 12 hours. The busiest day for printing has been Thursdays. 3D print job pricing is based on a cost-recovery model where equipment life, based on manufacturer projections, and cost of supplies (in particular, filament) are taken into consideration. A cost-estimating tool is included on the Innovation Space website to help users gauge potential costs of jobs. The cost is $3.00 for the first hour and $1.00 for each subsequent hour of printing. There is no charge for items that fail to print properly.

Fifty-four classes in 3D printing and modeling had a total of 209 attendees. Sixty-two local high school students from a university outreach program attended orientation sessions. Many early 3D print jobs were learning experiences, and users printed small, simple items, such as key chains. Use has become more sophisticated. Faculty have assigned 3D printing projects to students in the Physical Therapy program. Physicians from the University of Maryland's Shock Trauma Center have used the 3D printers to develop clinical models for more precise repair of facial bone injuries. Expensive lab equipment has been repaired with 3D-printed parts. Use of the space is limited only by the user's imagination.

NEXT STEPS

At a mid-2016 meeting of the university's Facilities Master Planning Committee, a participant remarked that the HS/HSL was the only place at the university where there were opportunities to try new technologies. Participants affirmed the importance of the library's neutrality and centrality to the student experience. It has been observed that the HS/HSL is a nexus for student entrepreneurial activities and experiential learning and that there should be more spaces like the Innovation Space and the Presentation Practice Studio created.

In addition to service on the Facilities Master Planning Committee, the executive director served on the university's Strategic Plan Steering Committee, which updated the university's strategic plan for 2017–2021 (http://www.umaryland.edu/about-umb/strategic-plan). This afforded the executive director an opportunity to assess future needs and influence future directions of the university. The new strategic plan focuses squarely on student success, excellence in research and teaching/learning, and an ongoing emphasis on collaboration and interprofessionalism. There is a synchronicity between the new strategic plan and initial conversations regarding the facilities master plan, to be completed in 2017. These initial facilities discussions have teased out the need for more collaborative, experimental, and even entrepreneurial incubator-type spaces. Currently, many of these needs are being met in the HS/HSL's Presentation Practice Studio and the Innovation Space.

What does this mean for the future of space planning efforts within the HS/HSL? There continues to be space availability due to fewer print collections. However, many of these spaces may not be optimal for the types of use envisioned. Is it possible to have disseminated innovation activities? Currently, the Presentation Practice Studio and the Innovation Space are easily accessible on the main and second floors of the library, with easy access to the staff supporting them. Stacks are located on the third through fifth floors, some distance from user support.

In the near term, the focus will be on evolving the services emanating from these spaces. There have been conversations about stand-alone media-editing workstations to extend the reach of some of the services of the Presentation Practice Studio. A new videoconferencing room will be located next to the Presentation Practice Studio, expanding that capability from the current two videoconferencing rooms to three. There will be serious investigation of placing videoconferencing capabilities in the largest, 40-seat library conference room, adding much-needed large-group space to the overall university inventory.

An expansion of the Innovation Space may include data visualization support and display walls. A poster printing service based on a cost-recovery model was launched in September 2016. A strategic evaluation of the layout of the main floor of the library is under way. This evaluation will essentially segment the floor into innovation, technology/collaboration, and study zones. Programmatically, collaborating with outside groups and other universities to offer higher-level courses, hackathons, and internships, as an extension of the services related to the space, are in discussion. The university does not have a computer science program. In order to expand expertise, partnerships are being explored with University System of Maryland universities and other institutions of higher learning. There are also plans for the university's informatics incubator to move into an upper-level library space. The synergies of having these types of capabilities in one building are being explored, and this could lead to exciting partnerships and developments.

Perhaps the most challenging next step regarding these two areas is developing plans for the sustainability of these spaces. Engagement with the university community and promotion of the library's expertise, resources, and facilities is constant. Ongoing environmental scans and need assessments regarding the usefulness and applicability of these services are necessary. Sustainability involves hiring and retaining the staff necessary to provide the highest level of support and service. These staff have different expertise than traditional library staff, requiring different types of professional development. They may have differing expectations regarding the work environment as well.

Technology also has to be sustained, upgraded, improved, and replaced, requiring an ongoing financial commitment to the spaces. To date, donors have been generous in support of these initiatives. Can an endowment be established to ensure ongoing support? Currently, the Presentation Practice Studio and the Innovation Space operate on a materials cost-recovery basis. Can these services be expanded to users, perhaps in bioparks or small companies, outside of the university as a cost center? Should there be charges for inclusion of these services and staff expertise in grant submissions or for online course development? These questions need to be considered.

LESSONS LEARNED

With every one of these projects, lessons are learned, and there are teachable moments for everyone involved. Although some of the following 11 lessons are more theoretical than practical, all were fundamental to the success of these projects:

1. Constantly scan the horizon for new ideas. Read in diverse subject areas beyond your traditional ones. Look for commonalities and trends. Try to imagine how they would work in your environment or if they would even be appropriate. Can you meet a need? Put things on a back burner if timing is not right. Tune into what is important to your institution.
2. Practice agility. It is important to be able to move quickly when something feels right.
3. Be prepared to advocate for and defend the project. Learn all you can. Have data to show that the project is important and needed. There will be naysayers and doubters. Be prepared.
4. Have conviction and belief that the project is meaningful and will succeed.
5. Plan, plan, plan. The more you plan, the more sound the resulting foundation. If the planning raises doubts, uncovers obstacles, or identifies challenges, this will improve your project or cause a rethinking of it altogether. Assemble an initial project team and spend no more than six months examining the issue and developing solutions and recommendations.
6. Practice what you preach and collaborate. Get outside opinions about your plan. Develop an advisory committee. Listen to the input of staff involved.

7. Appoint someone the project manager or decision maker. Eventually, someone has to have responsibility for making key decisions and moving the project forward, even if it is deciding on paint colors. Decide how much you want to be involved and how much you trust others.
8. Flexibility is key. Things do not always go as planned. There are delays. Ideas need to be reevaluated.
9. Unabashedly promote the project. Mention it in meetings; put it on agendas. Write articles and press releases. Post flyers. Invite people in to see the construction as it unfolds. Hold a grand opening. Find advocates in the community; arm them with information so that they can help with the promotion.
10. Assess and evaluate the project. Decide what constitutes success ahead of time and measure it. Collect anecdotes as well as data.
11. Know when to let go. Sometimes projects just do not work out. Develop the attitude that you learn as much from failure as you do success.

Good luck!

NOTES

1. Frieda O. Weise and M. J. Tooey, "The Health Sciences and Human Services Library: 'This Is One Sweet Library,'" *Bulletin of the Medical Library Association* 87, no. 2 (1999): 170–77.

2. M. J. Tooey, "Renovated, Repurposed, and Still 'One Sweet Library': A Case Study on Loss of Space from the Health Sciences and Human Services Library, University of Maryland, Baltimore," *Journal of the Medical Library Association* 98, no. 1 (2010): 40–43, http://doi.org/10.3163/1536-5050.98.1.014.

3. Brad Gerhart, Ryan L. Harris, Janice E. Kelly, Tierney Lyons, Alexa A. Mayo, and Paula G. Raimondo, "Collaboration Space: A White Paper," http://hdl.handle.net/10713/435 (accessed June 27, 2016).

4. Ibid., 5.

5. Ibid., Appendix B, 13–14.

6. 3D Printing Industry, "The Free Beginner's Guide: 02—History of 3D Printing," 2016, http://3dprintingindustry.com/3d-printing-basics-free-beginners-guide/history (accessed June 24, 2016).

7. Steven Leckart, "How 3-D Printing Body Parts Will Revolutionize Medicine," *Popular Science*, August 6, 2013, http://www.popsci.com/science/article/2013-07/how-3-d-printing-body-parts-will-revolutionize-medicine (accessed June 24, 2016).

8. Bohyun Kim, Everly Brown, Aphrodite Bodycomb, and Thom Pinho, *Makerspace Task Force Report*, 2014, University of Maryland, Founding Campus Digital Archive, http://hdl.handle.net/10713/4634 (accessed June 27, 2016).

Chapter Six

Gary L. Crocker Innovation and Design Laboratory, University of Utah

Jean P. Shipman and Timothy Pickett

An important phase of innovation is prototyping. This is where the ideation and concept development components of innovation get translated into reality—where vision and design get realized as a sample of the product is made (proof of concept). Several types of prototyping are described herein. This chapter also highlights the experience at the University of Utah (U of U), where the Center for Medical Innovation (CMI) partnered with the Spencer S. Eccles Health Sciences Library (EHSL) to install a high-fidelity fabrication laboratory, retrofitting an EHSL mailroom space. Guidance as to what equipment, functionality, staffing, furniture, regulations and preparations, funding sources, and lessons learned is offered to those considering such facilities within their own universities.

WHAT IS PROTOTYPING?

Prototyping, as defined by *Merriam-Webster's Collegiate Dictionary* (http://www.merriam-webster.com/dictionary/prototype), is "an original or first model of something from which other forms are copied or developed later." There are three main variations of prototyping: additive, subtractive, and electrical. *Additive prototyping* is building a product, layer by layer. This is the common technique used by many 3D printers. Materials come in a large variety of components, including all kinds of plastics. Products can be created that have objects within objects and interlocking components. *Subtractive prototyping* is taking a solid and removing material from it to form a shape or final product. One vendor that sells subtractive equipment (e.g., CNC mills) is a company called Tormach. Another company, Proto Labs, will actually take your design and fabricate it for you so that you do not need to invest in expensive equipment. Originally, subtractive prototyping techniques were more involved and required operators with extensive experience versus typical additive prototyping techniques due to the advanced computer coding techniques that were required. However, technology has advanced to the point where the computer coding can be done automatically to eliminate this drawback. *Electrical prototyping* is where the electrical aspects of the product are incorporated to determine its functionality according to specifications. Each type of prototyping requires different kinds of equipment, expertise, and refinement. All of these types of prototyping should be included in a prototyping laboratory, with the goal of being able to create proof of concept or beta prototypes.

BACKGROUND

In early 2014, the executive director of U of U's CMI approached the EHSL director to inquire if any space within the EHSL could be assigned to establish a prototyping laboratory to be used by students, faculty, innovators, industry members, and community individuals. Such a lab would be used primarily for competitions such as Bench to Bedside (B-2-B) but would be open to anyone needing prototyping assistance and facilities.

As many libraries were putting in makerspaces, the EHSL director agreed that a collaboration with CMI to create a prototyping lab would be a positive new asset for EHSL and decided to rework the EHSL's mailroom (540 square feet) for that purpose. The mailroom had been used at one time as a mailroom but most recently as a storage facility for photocopy paper, construction materials, and other items. This space had concrete block walls, a cement floor, a small sink, numerous electrical outlets, an outside door that connects to the EHSL loading dock, and electrical control switches for lights for the lower level of EHSL. The CMI paid to have flooring added to cover the cement, and the EHSL got to work on clearing out the space, including removal of the sink and relocation of the electrical control switches. CMI ordered furniture, tools, cabinets, and equipment and proceeded to hire a director of engineering who started in November 2014. This director has expanded the functionality of the lab to support and educate students, faculty, and industry partners about product design, specifications, prototyping verification testing, usability testing, and the entire product development process.

THE GARY L. CROCKER INNOVATION AND DESIGN LABORATORY

The need for more equipment, staff, and functionality soon arose, and through generous donations from Utah entrepreneur, philanthropist, CMI community partner Gary L. Crocker (http://www.crockerventures.com/about/gary-l-crocker) and the Sorenson Legacy Foundation (http://sorensonlegacyfoundation.org), this expansion need is being addressed. Tim Pickett, the CMI director of engineering, has been working to implement these new additions and has increased the footprint of the Crocker Lab (figure 6.1) by installing equipment within a Skills Center, also located on the lower level (Garden Level) of the EHSL. He obtained a sink that was installed where the previous sink had resided, as plumbing was already in place. Water and drainage are two main requirements for a prototyping lab. Details about the Crocker Lab are provided next.

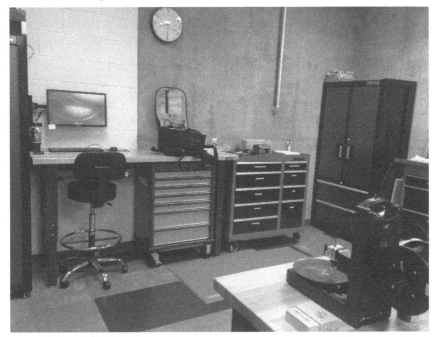

Figure 6.1. Crocker Lab.

Purpose

The Crocker Lab is designed to help students and professionals design and innovate medical products in every stage of development. Engineers and graduate student fellows with expertise in electrical engineering, mechanical engineering, manufacturing, usability, human factors design, and industrial design are able to assist anyone with complex issues.

The Crocker Lab is open to students, faculty, the U of U community, and industries outside of the U of U to prototype medical devices, verify testing of devices with equipment calibrated to NIST standards, perform usability studies, and get consultations from industry and academic professionals on specialized topics.

Staffing

The Crocker Lab has one full-time U of U employee: the director of engineering. Student fellows are hired to offer a range of expertise to users. These fellows come from the bioengineering, mechanical, and electrical engineering departments. Fellows share their expertise with Crocker Lab users, but they also benefit from this employment as they gain real-life experience that is adjustable with their class schedules, receive some compensation, and get to know the movers and shakers within the industry and the community. They also are able to document this employment experience on their résumés as they seek employment postgraduation.

Large Equipment

The Crocker Lab houses a variety of equipment to meet the different types of prototyping, including additive, subtractive, electrical, and rapid.

Additive Prototyping

One way to explain what additive prototyping or manufacturing is about is to compare it to a body scan, where a body is examined layer by layer or one slice at a time to build a composite picture of the whole. In the library world, additive prototyping can be likened to the Visual Human Project (https://www.nlm.nih.gov/research/visible/visible_human.html), where an entire anatomical dissection of two human beings was achieved by stacking thin slices of the bodies on top of one another until the entire bodies were replicated. 3D printers that provide this kind of composite-building prototypes include the following:

   a. Fused deposition modeling (FDM)
   This 3D printer operates like a hot-glue gun that exudes plastic into an x-y-z coordinate system using a variety of different materials. There are many types available that are relatively inexpensive.
   b. Stereolithography (SLA)
   This equipment uses a liquid resin that is cured with a precisely aimed laser. It is more expensive and often a hassle to work with, as there is a lot

of maintenance involved and use is messy. A special resin limits what choices one has with material options, but the material is easy to purchase.

c. PolyJet

A PolyJet is a big 3D printer that has the highest resolution and uses photopolymer resin that is cured by ultraviolet light as it is sprayed. This 3D printer is expensive and has a limited material choice that has to be purchased directly from a printer manufacturer, such as ProtoCAM or Stratasys. Maintenance on PolyJets is an issue—equipment requires weekly maintenance and also after every print job, requiring staff time. One can obtain yearly maintenance policies. This printer can use all kinds of materials, including metal, food, tissue, and so on, including a mixture, and can create multicolor and pliable products.

Subtractive Prototyping

With this type of prototyping, equipment cuts away pieces from a solid. Such equipment includes items such as band saws, drill presses, belt or disc sanders, grinders, CNC mills (figure 6.2), and lathes that handle metal, wood, and other kinds of materials. The Crocker Lab has all of these kinds of subtractive prototyping tools and offers assistance with their proper use, including safety measures to follow when using them.

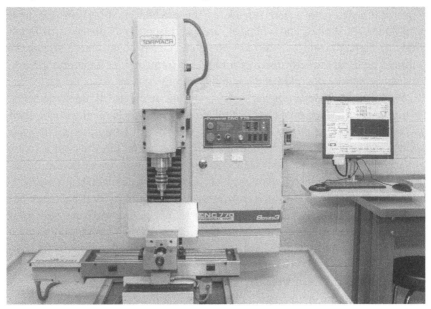

Figure 6.2. CNC Mill.

Electrical Prototyping

Electrical prototyping requires very specific equipment to run conduits for electricity and to measure input and outputs. The main piece of equipment for this kind of prototyping consists of Arduino microcontrollers that have controlled data inputs and outputs to make measurements. "Arduino is an open-source electronics prototyping platform based on flexible, easy-to-use hardware and software. It's intended for artists, designers, hobbyists, and anyone interested in creating interactive objects or environments."[1] The Crocker Lab also has soldering irons, analog circuits, and a variety of sensors to collect different kinds of data inputs.

Rapid Prototyping

With any kind of rapid prototyping described in this chapter, small quantities of products are created and tested before the concept is scaled and large volumes are produced. One of the reasons for developing prototypes is to ensure that the product concepts function as intended and are safe to use. To determine functionality, several kinds of measurements need to occur within a fabrication laboratory, including electrical, mechanical, and temperature. Specific equipment is needed for each measurement type, as is outlined next.

Electrical Measurements In order to take electrical measurements, several pieces of equipment are needed, such as a direct current (DC) voltage power supply, a volt meter (measures three different things in one unit: amplitude, resistance, and voltage), an oscilloscope (to measure different kinds of signals), and a function generator (i.e., used to create signals or multiple types of waveforms over a range of frequencies). Function generators can be battery operated and come in a variety of sizes.

Mechanical Measurements In addition to electrical measurements, prototype creation requires a set of tools to measure mechanical aspects, such as length, height, width, weight, radius, and so on. Tools that assist with taking precise measurements include a weight scale and set of weights, rulers of variable lengths (need both metric and inches [millimeter or half millimeter, up to 1/64 of an inch]), tape measures, levels, digital calipers to measure interior and exterior dimensions of products (tip: get one that switches between inches and millimeters), and body scanners that measure and print body replicas or selected parts via 3D printers. Body scanners come with different functionality and resolutions, depending on the accompanying software.

Temperature Measurements Prototype generation also includes needing to measure the temperature of various components. Equipment that is needed to do such measurements include thermocouples (multiple kinds exist that withstand different temperatures and offer a variety of protection tube diameters), ceramic or polymer thermistors (i.e., resistors used to sense temperatures), and temperature-sensing bulbs.

Laser Cutters

Laser cutters create two-dimensional items or prototypes. There are two main kinds of laser cutters: CO_2 and fiber. CO_2 laser cutters work with acrylic and various polymer materials and also can be used to do engraving. These cutters are great for medical device prototyping. The second kind, a fiber laser, is used with polymers and metals and tends to be more expensive than CO_2 laser cutters.

Injection Molding

> Injection moulding is a manufacturing process for producing parts by injecting material into a mould. Injection moulding can be performed with a host of materials, including metals, (for which the process is called diecasting), glasses, elastomers, confections, and most commonly thermoplastic and thermosetting polymers. Material for the part is fed into a heated barrel, mixed, and forced into a mould cavity, where it cools and hardens to the configuration of the cavity. . . . Injection moulding machines consist of a material hopper, an injection ram or screw-type plunger, and a heating unit.[2]

The beauty of injection molding is that it can be used to create high volumes of similar products and is commonly used to manufacture plastic items.

Small Equipment and Supplies

In addition to the large equipment, there is a need for many different types of smaller equipment and supplies. Tools such as tweezers, wrenches (including a ratcheting wrench set and Allen wrenches), screwdrivers (assortment of sizes and types), hammers, mallets, nuts, bolts, washers, screws, cap screws, pliers, saws (hack and band), SAE and metric fasteners (an assortment), wires, wire cutters and strippers, hand drills, Dremel tools, vices, hot plates, pancake air compressors, sanders, and shop vacuums are necessary elements of a fabrication laboratory.

Supplies and chemicals that should be on hand include batteries, sandpaper, steel wool, a sewing kit, different adhesives (e.g., epoxies, superglue, Elmer's, and wood glues), different kinds of tape (e.g., duct, masking, cellophane, electrical, painter's, and aluminum), acetone, isopropyl alcohol, and a variety of raw materials.

Safety Supplies

Safety supplies and equipment should also be a part of a fabrication lab. Include such items as the following: first-aid kit, sharps containers, fire extinguishers of multiple types, safety googles, gloves, shop coats, guards for protection with projectile equipment, and chemical storage cabinets (nonflammable). Individuals working in the Crocker Lab receive training on how to use the equipment properly to avoid safety concerns, and they are taught what equipment requires safety supplies, such as goggles and gloves.

Furniture

Depending on the space size, different types of furniture should be acquired, including workbenches, stools, display cases, and cabinets for tools, books, and supplies. Do not forget a trophy case or two to showcase competition awards and actual prototypes.

Additional Facility Considerations

There are several facility considerations to take into account when establishing a prototyping laboratory. Electrical utilities must be at their maximum in order to support the requirements of a prototyping lab. Be sure to have the proper exhaust hoods, fans, and outlets, as some of the materials, when heated, can release toxic fumes, and particles can be dispersed into the air from the different printers and mills. Noise can also be an issue when operating the equipment, so sound hoods are highly recommended, especially if the laboratory is located within a library.

Software

Computer-aided engineering (CAE) is an indispensable tool for prototyping. A digital 3D design is the first prototype. Three-dimensional designs are created using computer-aided design (CAD) software. Some examples of CAD software are Solidworks, Solid Edge, and Autodesk Inventor. The CMI has partnered with the U of U College of Engineering to obtain a number of licenses, as it has an educational license agreement for Solidworks. CAE software can be very expensive; funding partnerships should be explored, or consider using open-source versions, such as FreeCad.

OTHER ISSUES TO CONSIDER

Now that you know what equipment, supplies, furniture, and safety items are needed to properly outfit a fabrication laboratory, there are many other issues to consider and decide how to handle. These include governance of the space (which becomes more complicated if the lab is contained within a multipurpose building), egress and security, parking, and sustainable funding.

Governance

Rules about proper laboratory use need to be established and posted so that users are aware of whom they need to contact if issues arise with using the equipment and facilities. Maintenance of the equipment needs to be outlined and followed to ensure that proper operations ensue with repeated use. Being housed within the EHSL requires that the Crocker Lab follow additional governance considerations, such as how to order facility repairs and improvements, building security guidelines and how to obtain after-hours building access, knowing proper disaster procedures (i.e., phone trees, reporting spots on campus, having on hand required disaster supplies, and so on), how to receive packages, U.S. and campus mail, and expectations of use of shared café and lounge facilities. Signage also has to be approved by the EHSL, as they remain the facilities managers of the library building. Office and conference room keys also are issued by EHSL, so proper university request protocols need to be followed.

Egress and Security

Anyone can gain access to the Crocker Lab during set operational hours, currently set at 9:00 a.m. to 2:00 p.m., Monday through Friday. Innovators who are working on products for the B-2-B competition are granted access 24/7. Others can make special arrangements for after-hours access with the director of engineering. With egress needs of 24/7, the EHSL had to remove the building's intrusion alarm and include designated individuals as having access rights within the U of U's security card system. With the loading dock door opening right into the Crocker Lab, individuals with card system permissions can enter and exit via this set of metal double doors after EHSL operating hours. The main door to the Crocker Lab opens into the Garden Level foyer and is secured with keys issued by the EHSL. The doors are wooden double doors; they host a sign posting the Crocker Lab's general operational hours. Having double doors is great, as large equipment can be moved in and out of the space as needed.

Parking

One egress issue that remains a problem is parking. Universities are notorious for having insufficient parking, and, unfortunately, neither the EHSL nor the CMI have dedicated parking spaces for their clients. Available parking is metered. Individuals visiting the Crocker Lab can park in a visitors' lot that is located three blocks away and is a fee-based lot; thus, parking validations are a hidden cost for CMI, as they cover these costs to encourage people to come to the Crocker Lab.

Sustainable Funding

Initial support came from the U of U vice president for health sciences to outfit the Crocker Lab space. As more equipment was needed, as well as additional staffing, donations were received from Gary L. Crocker and the Sorenson Legacy Foundation. Material costs are reimbursed by B-2-B team stipends ($500 per competition per team) and otherwise directly by users per issued quotes and invoices. Grants from foundations and extramural programs are being sought to further expand Crocker Lab facilities. Additional donations are being obtained from a variety of sources.

LESSONS LEARNED AND TIPS

Starting a fabrication lab is not for the faint of heart. It takes time, money, and tenacity. Many lessons have been learned with the installation of the Crocker Lab, and, inevitably, there will be more as the Crocker Lab continues to grow and thrive. A key recommendation is to try to outfit a lab with the proper equipment, exhaust, and plumbing infrastructures from the beginning, as this will save on construction or remodeling costs and ensure proper operations. Strategize about future growth from the start and predetermine how additional space can be acquired as the lab expands in function and usage. Always consider how to raise money for a lab; talk with others within your university to cultivate potential donor relations and express case needs to your development office.

Some practical tips offered include the following:

1. All labs need water, exhaust outlets, and drainage. Be sure these are included in your lab space from the start. Ensure that a glove box or water jet comes with your sink to be able to remove materials from additive prototypes.
2. When purchasing equipment, ask for educational discounts and know that a lot of equipment is available only through educational retailers.
3. One can never have enough storage. Consider storage for supplies, materials, devices, tools, and so on. Add shelves to store such items.
4. Put a *quality* system in place from the beginning.
5. Establish success metrics at the start and assess the lab's impact regularly. Promote the impact to others throughout the institution and to donors.
6. Never call scraps of materials *scraps*; instead, call such materials *remnant prototyping*. These materials can assist with the ideation stage of

innovation, as they provide the opportunity to play with design and ideas to create rough prototypes.
7. Collect expired medical devices and supplies from your hospital, as they need to resupply their inventories per set time standards. Decommissioned devices can be used by innovators to emulate designs, to explore functionalities, and to glean what has been already produced and is available on the market.
8. Have an outside entrance to a lab if it is placed within a multipurpose building. Innovators need to work around the clock, and such direct egress enables 24-hour access.
9. Look at universities that have existing fabrication laboratories to see how they operate and are structured. Known universities with such facilities include MIT, Stanford University, UCLA, and the University of Minnesota. Several other institutions have such facilities as well, such as Cleveland Clinic and Mayo Clinic.
10. There are several websites that can be monitored to stay current with the latest information related to fabrication or prototyping laboratories. Examples include YouTube, LLC, Instructables.com (http://www.instructables.com), GrabCAD.com (https://grabcad.com), and Thingiverse.com (http://www.thingiverse.com).

FUTURE

Plans are under way to build a new Discovery Center building that will include a prototyping laboratory that is bigger and more sophisticated than the Crocker Lab as far as infrastructure and functionality. This new lab will be co-located with a gaming laboratory, a simulation center, a showcase display area, conference rooms, and individual offices. The Discovery Center is currently under design to be built within the next four years. It will be part of the Health Sciences Transformation complex of the U of U. Meanwhile, the Crocker Lab will continue to partner with other fabrication labs available within the Innovation Ecosystem at the U of U, including a new 400-dorm innovation center that opened in September 2016—the Lassonde Studio, a division of the David Eccles School of Business.

NOTES

1. Arduino, https://www.arduino.cc (accessed October 12, 2016).
2. Robert H. Todd, Dell K. Allen, and Leo Alting, *Manufacturing Processes Reference Guide* (Norwalk, CT: Industrial Press, 1994).

Chapter Seven

Library Makerspace Programs

Bringing Together Space, Services, and Staffing

Elliot Felix and David Woodbury

EXPANDING THE LIBRARY ROLE

Libraries have seen a shift in their roles as warehouses of books and journals to providers of spaces and services in support of teaching and research needs on campuses. This shift began with an expansion of access (from closed to open stacks, 24 hours) and has led to the more recent developments of spaces, such as *learning commons*, with expert help, collaboration spaces, and digital tools close at hand. Makerspaces are the latest of new specialty spaces in libraries that provide areas for guided instruction, informal use, and scholarly creation in new media. Why makerspaces in libraries? Because libraries provide opportunities to support making in a departmentally neutral (and often central) location, to connect digital and physical making, to draw from information and expertise within the library, and to provide a platform to showcase what is made to inspire others. Makerspaces in libraries offer democratized access to tools and technologies that serve to maximize benefits for large numbers of people. Rather than hiding in departmental labs, these tools are exposed for all to use within a library setting regardless of discipline. They also offer space for messier activities not traditionally supported within library walls. These spaces often serve to spark interdisciplinary connections, as they attract a diversity of users within the space.

Libraries can also provide online and print resources that tie to the skill sets needed for making 3D designs, programming, and electronics. What is made in the library can also benefit from library staff expertise, such as expertise for business or patent research as part of a developing product or service. Audio and video creation spaces can be connected through adjacencies or programmatically to makerspaces. Libraries also have a unique opportunity to connect making programs to other areas of strength, such as creating an online presence for the maker projects. Makerspaces provide opportunities for institutions to introduce *learning by doing* activities as well as to design thinking approaches to problem solving.

In this chapter, we draw from our experience to advocate for makerspaces in libraries, starting with why and designing not simply a maker *space* but rather a maker *program*. We then provide a methodology for holistically thinking through the spaces, services, and staffing that make up such a program. Finally, we identify the common challenges and opportunities for the implementation of maker programs.

STARTING WITH WHY

Makerspaces in libraries do not follow a single archetype and are implemented with varying forms of spaces, services, and tools. Because of this, institutions should start with specific institutional and programmatic goals and recognize that rather than simply a maker *space*, they will likely need what we call a maker program—a holistic approach to providing spaces, services, and staffing to support making and makers. Many well-intentioned makerspaces start with the purchase (or gift) of a 3D printer without thinking ahead to (or securing funding for) ongoing support or programming. Instead, makerspaces can model the design paradigm so crucial for the projects they intend to support: developing the *why* (the goals of the space) with users, implementing a program, and then iterating on that program. The process of planning the program can also be used to build interest and support by engaging the university community.

Universities are likely to have many spaces in a variety of departments that incorporate making. Makerspaces in libraries coexist with all of these spaces by providing a neutral location for all on campus regardless of discipline and by serving as an on-ramp to more specialized areas. The library can also fill in gaps from other programs by providing greater access and visibility for projects, often creating interdisciplinary connections that are not as possible in a department-oriented space. The library serves as a center for these disparate groups to help formulate a maker community hub. The visibility of the library over hidden spaces in departments, as well as the exhibit expertise of librarians, provides an opportunity for the library to showcase maker projects created at the library and elsewhere. A curated group of interactive, hands-on maker projects can serve to educate and inspire library visitors.

A successful library makerspace program can be as simple as inexpensive kits that are checked out from a circulation desk or as complicated as a large dedicated space with specialized heating, ventilating, and air-conditioning and industrial equipment. The goals can range from introducing students to making and digital literacies to supporting small manufacturing and entrepreneurship. The space should not be defined by equipment alone. Equally important is how groups of users will be supported within the space.

As libraries define their goals for creating a new makerspace or enhancing an existing one, they will be faced with a variety of questions: Can the space accommodate single users as well as groups of people for classes or workshops? How and where will workshops and other programs be held? What equipment will be provided? Will mediated access to equipment be necessary (i.e., for safety and security reasons), and how will it be implemented? Who will staff the space, and who will repair equipment? Faced with these questions, a systematic approach to stepping through the different dimensions of a makerspace program is recommended to figure out what might work: trying it out, assessing, and then learning and adjusting as needed.

HOW TO DESIGN A MAKERSPACE PROGRAM

Because libraries can support making—and affiliated activities such as creativity, innovation, and entrepreneurship—in a variety of ways, what is needed is a methodology that helps avoid common mistakes, such as leading with a solution prior to identifying programmatic goals or forgetting that providing maker *spaces* often means that you are also providing maker *services*. Our method for designing your makerspace program is to use a kind of taxonomy of makerspaces that can serve as a menu as you plan. Institutions should consider their programmatic goals, what spaces are available, what services will be offered, how these spaces and services will be staffed, what equipment will be provided, the way the space and equipment will be accessed, and the special considerations associated with each type of makerspace. Types include (organized generally in ascending order of cost, complexity, and risk) handheld equipment loan and events, designated space with bring your own device (BYOD), dedicated unstaffed makerspace with small equipment, staffed makerspace with small equipment, and staffed makerspace with large equipment (see table 7.1).

Table 7.1. Types of Makerspaces.

	Handheld equipment loan and events	Designated space with equipment BYOD	Dedicated unstaffed makerspace with small equipment	Staffed makerspace with small equipment	Staffed makerspace with small equipment	Staffed makerspace with large equipment
Example of this category:	The Labs at Carnegie Library of Pittsburgh	University of Nebraska Omaha "Messy Rooms"	University of Mary Washington Thinklabs	North Carolina State University D. H. Hill Library		University of North Carolina at Chapel Hill BEAM
Programmatic Goals	Assess interest, promote making, develop new literacies	Support nascent maker/entrepreneurship community who can self-manage	Support emerging maker community who can self-manage	Support growing maker community who need support based on scale of interest or complexity of projects		Support mature maker community who create complex physical objects in a wide variety of materials
Space	No dedicated space; pop-up events held prominent places (lobby, classrooms)	Designated space, such as a studio with raw durable furniture and finishes	Dedicated space, including shared layout/fabrication space, tabletop equipment, and material storage	Dedicated space, including shared layout/fabrication space, tabletop equipment, material storage, student project storage, utility sink, and staff space		Dedicated space, including shared layout/fabrication space, tabletop and large equipment, material storage, student project storage, loading dock access, hazardous waste disposal, and staff space
Services	Equipment lending, skills-based tutorials, and sharing/showcasing sessions	Facilities maintenance	Facilities maintenance and equipment install and maintenance	Facilities, equipment lending, equipment install and maintenance, skills-based tutorials, and		Facilities; equipment lending; equipment install, maintenance, and supervision; accepting payments;

Library Makerspace Programs 71

	Handheld equipment loan and events	*Designated space with equipment BYOD*	*Dedicated unstaffed makerspace with small equipment*	*Staffed makerspace with small equipment*	*Staffed makerspace with large equipment*
				sharing/showcasing sessions; may accept payments	skills-based tutorials; sharing/showcasing sessions; and large/high-risk equipment certification
Staffing	Community organizer with technical expertise	No dedicated staff but may be supported by student club or volunteers	No dedicated staff but may be supported by student club or volunteers	Dedicated staff during hours of operation	Dedicated staff during hours that large equipment is operating
Equipment	Arduino kits and miscellaneous prototyping supplies (paper, glue, wood, etc.)	None	Electronic kits, paper cutter, vinyl cutter, 2D printer, consumer 3D printer(s), whiteboard(s), and projector/display	Paper cutter, vinyl cutter, 2D printer, consumer 3D printers, sewing machine(s), soldering irons, small CNC machine(s), whiteboard(s), consumable supplies, and projector/display	Professional 3D printing, laser cutter, large-format printer, large CNC machine, welding equipment, power saws/drills, electronic lab bench, whiteboard(s), and projector/display
Access	Equipment open to all; events either drop-in or by advanced sign-up	Open to all during library hours	Open to all during library hours	Open to all but may have fewer open hours, depending on staffing	Open to all but may have fewer open hours, depending on large-equipment access

	Handheld equipment loan and events	Designated space with equipment BYOD	Dedicated unstaffed makerspace with small equipment	Staffed makerspace with small equipment	Staffed makerspace with large equipment
Special Considerations	Marketing, signage, online sign-ups, and online showcase for work	Relationship building with maker community to enable self-management	Relationship building with maker community to enable self-management	Managing access to supplies (free vs. fee) and project storage	Equipment use may have different hours than space use to increase access

COMMON OPERATIONAL CHALLENGES FOR MAKERSPACE PROGRAMS

Makerspace programs need champions to lead programs and also need technicians to fix and maintain equipment. Staff are also needed to mediate access to the more dangerous equipment or to sell materials or outputs from equipment such as 3D printers. These responsibilities are often accomplished by the same person. Many programs rely on student workers or volunteers to extend work and services, but this means there is supervisory overhead.

Initial access to the makerspaces is often dictated either by the hours of operation from a staffing perspective, by the safety programs for the equipment being used, or the ability to handle monetary transactions if based on a usage fee model. Many spaces simply require an initial safety overview and registration and then provide open access through a card swipe or sign-in process. Spaces with very dangerous machinery, such as wood or welding shops, have additional limitations on the use of this equipment with individual certifications (e.g., UNC BEAM, Thinkbox, and BLDG 610).

Makerspace owners may want to sell materials or establish a financial model that requires payment for use of all or some of the equipment. However, accepting payments can be a challenge for libraries. Incorporating an appropriate secure area for transactions, salable material storage, and paperwork could be necessary. Some makerspaces forgo money handling and create self-service options with vending machines stocked with consumable materials, such as electronics, filament, and so on. An example of this is the Artisan's Asylum Hackerspace.

COMMON FACILITIES CHALLENGES FOR MAKERSPACE PROGRAMS

Basic makerspace programs need basic facilities to get started. As the equipment increases in complexity and cost, so do the infrastructure, health, and safety issues. Depending on the activities and the equipment, all that may be needed is easy access to power. However, as equipment gets larger and a variety of materials are used, these elements impact space requirements. Big equipment needs more dedicated power. Laser cutters produce fumes that should be vented externally. 3D printers create fine particulate matter that requires adequate ventilation. CNC and woodworking machines produce a lot of waste.

Consulting with health and safety experts on the types of materials and the amount of equipment used is important to consider early in a project. Likewise, determining which safety measures are needed, such as eyewash stations and the use of heat detectors rather than smoke detectors where fumes are likely to be created, is also a consideration. Attention should be given to the networking infrastructure and setup. Makerspaces are likely to use a lot of wireless bandwidth. Most campuses have restricted access points that may complicate the creation of *Internet of Things* projects that rely on wireless to connect new types of objects to the Internet. Libraries may work with information technology to create special access points with *white lists* for these types of experiments.

Makerspaces also need to accommodate the particular model of access. Is there a locked door? Do staff sit at the entrance? What can visitors see and do when they do not have access? Many spaces use large windows or glass walls to safely expose the internal activities to an external audience. Are items allowed to be removed from the space? If so, how do these items get checked out to users. If it is okay to store projects and materials, lockers should be located nearby.

LOOKING AHEAD TO WHAT IS NEXT

Once you have thought holistically about the space, services, and staffing needed to create a successful makerspace program and dealt with the kinds of operational and facilities challenges identified, you will be faced with another big question: what is next? This could mean how the program will change based on the feedback given. This might mean thinking about how you could scale up, and this brings with it questions about whether to adopt a more centralized or distributed strategy, and, if you do have multiple locations, how will they relate? Will they all have the same offerings and prioritize consistency of space, services, and staffing, or will they vary based on the location, needs, or constraints? Another key question institutions typically face is, as their interest in and support for making grows, do they wish to broaden the base of users to include other disciplines that are not yet engaged? Do they want to invite in the external community? How should limited resources be allocated for the greatest impact?

With the inclusion of makerspaces within a library comes a responsibility not only to provide inspiring and effective spaces, services, and staffing for making but also to capitalize on the opportunity to reflect on what these activities mean for librarianship and your library's practices and processes more broadly. What will the impact be on collection development, on community engagement, on access and circulation, and on instruction? For example, what is the library's collecting responsibility for the scholarly products that come out of a makerspace program? How will you document, curate, and showcase the work? What can be learned from how you engage maker communities that can be applied to how you engage other communities of interest? What learnings from thinking holistically about creating not only a maker *space* but also a maker *program* can be applied more broadly within the library?

Now is an exciting time for libraries. More people are coming to them than ever before. People are creating more diverse things than ever before—papers, videos, apps, websites, performances, and prototypes, to name a few. With the right makerspace program, an iterative approach that continuously refines the program, and a reflective attitude, libraries and librarians can help users solve problems in new ways and demonstrate lasting value to their communities.

Chapter Eight

Interview with Roger Altizer, PhD, and José Zagal, PhD (March 2, 2016)

Jean P. Shipman and Barbara A. Ulmer

BACKGROUND

There was an unmistakable rise in activity level in the Spencer S. Eccles Health Sciences Library (EHSL) when the Therapeutic Games and Apps Lab became an occupant in early 2014. Better known as The GApp Lab, this unique entity is a collaboration between graduate students in the University of Utah's (U of U's) number one world–ranked Entertainment Arts and Engineering (EAE) Program, the Center for Medical Innovation (CMI), and the EHSL. These students are funded to create medically relevant video games and applications. They are guided by two faculty members, Roger Altizer, PhD, and José Zagal, PhD. Following is an interview with these two faculty members to illustrate how The GApp Lab came to be located within the EHSL (see figure 8.1), how The GApp Lab applies information to what is produced, and how they, as faculty, perceive how libraries can partner with gamers of all varieties to address information needs and preserve the intellectual property being generated by such individuals.

Interviewers: Tell us a little about yourselves—your backgrounds and training, how you got into gaming, and, in particular, how you became involved in medical gaming.

Roger Altizer: My name is Roger Altizer. I am the director of the Therapeutic Games and Apps Lab and the associate director and cofounder of the EAE Program. I am also an adjunct professor in the Department of Population Health Sciences, School of Medicine. I had the good fortune to be born in the San Francisco Bay area during the rise of games, so as a

Figure 8.1.   The GApp Lab.

youth, I became involved in testing new games at the mall and also had an uncle in the industry, so I was exposed to gaming through him. I had not originally seen myself as professionally being involved in gaming. I did my undergraduate degree in religion and philosophy and taught English as a second language for a while. After graduate school as a communications major, I became a gaming journalist and wrote for About.com, a *New York Times* website, for 11 years, which was my professional start in gaming. I started teaching gaming and became involved specifically with medical gaming in 2012 when I worked on a patient empowerment game for children fighting cancer that was fairly well accepted and garnered attention from the press and a journal article in *Science*. Through that, I was introduced to health sciences people, and the growing interest in medical gaming resulted in our deciding to start a lab. In 2014, we launched The GApp Lab as a collaboration between the EAE Program, the CMI, and the EHSL.

José Zagal: Hello, I am José Zagal. I am faculty with the EAE Program here at the U of U, and I am also associated with The GApp Lab. My background includes most recently a PhD in computer science. Before that, I was working in collaborative education and the learning sciences with game research work on the side. Prior to my PhD, I earned a master's degree in engineering and computer science because, in Chile, a computer science degree meant needing a degree in engineering. While working on my undergraduate degree in the late 1990s, I was fortunate to have the opportunity to work on a research project for one of my professors involving handheld console education games for first graders on the Gameboy platform for basic math and Spanish. I became interested in medical gaming when Roger pulled me into becoming involved in The GApp Lab. At first, I was skeptical of the benefits of health-related games due to the *edu-tainment* movement in the mid-1980s, much of which did not turn out to be beneficial. My work on the PhD level in part was in order to obtain a deeper understanding of why that happened. So in that sense, it has been interesting to participate in The GApp Lab to see what new things we can do and how technologies have evolved to lend to more successful products.

Interviewers: What is The GApp Lab, and how did it come to be established? How did you become involved with it? And how does it fit into the overall U of U organizational structure?

Roger Altizer: The GApp Lab is a place where faculty collaborate with students to produce software, focused on medicine but not exclusively. We are also involved in some learning games, such as those focused on critical thinking, some activism games, and some prototyping for early commercial use. We also do core research projects where, for example, we explore questions such as the ability of the Microsoft Band to measure stress, which is knowledge that can then be applied in gaming. All projects are collaborative and funded by various mechanisms, including grants and internal money from the U of U promoting certain types of growth or research projects. Some are fulfillment of others' grants that are looking for specific sorts of collaboration partners. The GApp Lab is unique in that it is a U of U Main Campus and Health Sciences collaboration. The U of U has the EAE Program, the number one–ranked graduate student video gaming program in the world, and that program is collaborating with health sciences to do interesting development work and research.

The GApp Lab got started when we were doing more and more gaming types of projects in the EAE Program that involved faculty and staff. These projects were being sponsored by a variety of internal and external sources, such as the National Energy Foundation, who wanted to do an educational program for kids to learn about the Energy Safety Council. We decided a better way to organize this effort would be if we built a lab.

At the same time, the CMI executive director approached us and asked if we would be interested in collaborating on similar types of medical gaming projects. Initially, it was an exploratory effort; we were introduced to Jean Shipman and some other Health Sciences people. We ended up deciding to bring students and faculty interested in this sort of work to EHSL to take advantage of the good support and the proximity to the Health Sciences and to also have access to librarians and world-class researchers to help us do better work. We received a bit of seed money to help us get started. The EAE Program, CMI, and EHSL all donated staff time and equipment, and we started from there. This worked out extremely well; we soon had partners coming out of the walls and continue to have people coming by weekly asking what we do and how we do it. We now always have collaborators.

The senior vice president of Health Sciences lent her support from the beginning, so we continue to be gamers, meaning we are permitted to be loud and disruptive, maintaining the culture that supports the necessary creativity.

The GApp Lab fits into the overall U of U's structure through collaboration between the various contributing partners. Right now, The GApp Lab is still in its start-up phase and continues to receive funding from the EAE Program and Health Sciences, along with support from the CMI, EHSL, as well as other granting organizations. There are many departments that have a foothold in the lab and its partnerships.

José Zagal: I would like to add a further point of clarification on why the EAE Program was interested in becoming involved in The GApp Lab. If we go further back, one of the initial problems that the EAE Program experienced was that we wanted students to have practical professional experience, but the EAE Program grew to the point where they were not able to find enough internships. We considered internal programs where we could pay the students and have them work for us to gain the needed practical experience. We were looking for an internal collaborative structure to take advantage of economies of scale to be able to offer this experience to more students with fewer faculty needed. A facility like The GApp Lab met this need.

But a deeper reasoning for involvement in The GApp Lab was that outside the university walls, games have become ubiquitous, meaning it is our responsibility to make sure that students become aware of that ubiquity and that they therefore can work in all sorts of industries that use gamers, such as museums, companies with medical gaming interests, or educational organizations with interest in creating or evaluating educational gaming. We have noticed in the last few years, since our involvement in The GApp Lab, student focus has shifted from traditional gaming commercialization to interest in medical or educational gaming development. There have been success stories of students graduating and working for companies making educational and medical games. We have also seen students graduate and become entrepreneurs, starting their own companies developing health games. For example, one student has launched his own company to promote a game *Step Pets*, an incentivized pedometer-based game that allows you to earn levels where you are given virtual pets and their needed amenities based on how many steps you take daily. The GApp Lab has served to give students a broader range of interests as they have become aware of additional employment options.

I became involved in The GApp Lab because I did not have a lab of my own, and the consolidation sounded really interesting and made sense.

Interviewers: What is the most inspiring thing about working in The GApp Lab?

Roger Altizer: That has to be the sheer energy and optimism that happen due to the collisions. Collaborations can be difficult due to basic things like administration, accounting, and variations of perspectives on knowledge and how we learn. Because of the playful nature of games and the focus on the future, there is a high degree of optimism, along with some skepticism. Energy levels remain high when people work with others on things that make them excited. That is an important thing to feed.

José Zagal: Diversity of projects is vital for sources of inspiration. Not everyone is excited about everything, but with so many projects going on, there are always things happening that are exciting to everyone. That means you get to talk to people you would not talk to otherwise, and you get to hear about things you would not have without these collisions. There are different technologies, different problems being addressed, perspectives, contacts, etcetera.

In that regard, the name is a bit too narrow. What makes The GApp Lab unique compared to other labs is that other labs tend to focus on one narrow aspect of gaming, such as educational games for basic skills like math and science for grade school–age kids, but in The GApp Lab, we work on games ranging from . . .

Roger Altizer: . . . music education to spinal cord injury training games . . .

José Zagal: . . . to nurse intravenous training to learning how to draw pictures on the Web . . .

Roger Altizer: . . . to creating new medical imagery that is culturally sensitive . . .

José Zagal: . . . to helping kids plan a trip to the museum . . .

Roger Altizer: . . . to facilitating volunteers in India reporting malaria outbreaks . . .

José Zagal: . . . to assisting young adults make better decisions about health care insurance . . .

Roger Altizer: . . . to helping kids be less afraid of a visit to the optometrist.

Interviewers: Give an example of the most interesting innovation The GApp Lab has produced to date.

José Zagal: One of the most interesting to me was the Arches Insurance Company game, *Arches Saves Your Bacon*, a project that helped young adults understand the health care marketplace and understand that there is affordable health care. It also taught them why an annual physical is important.

Roger Altizer: I would have to choose the *Reflex Speed* project, which came about when a local entrepreneur wanted to challenge the notion that you cannot increase the reaction time of reflexes. We helped him with a project that was really interesting and allowed students to share their knowledge of engagement via software games and their expertise of software and hardware development through prototyping.

Interviewers: How important do you feel being within a library is to successful innovation development?

Roger Altizer: Being located in a health sciences library is a distinct advantage in many ways and offers a number of incredible opportunities, such as access to things like e-channel. One major challenge that exists for anyone developing software, particularly in an academic setting, is how to access information that can drive or inform project design. Inventors and software innovators are normally not domain experts. In our case, we work in so many different domains that it would be impossible to become experts in them all. There is a need to have access to both literature and the people that the EHSL has connected us to, many of whom are domain experts, and when they are not, they have the knowledge of how they can get us that information.

José Zagal: I think also the thing is that libraries in general are a neutral space within the academic community, so it becomes much easier to set up collaborations. If, for example, engineering is collaborating with law, at some point the question arises as to where they will establish offices and work spaces. Libraries are the Switzerland of academia. It also means that since librarians interface with everyone on campus, it makes the networking opportunities and collaboration possibilities that much richer.

Roger Altizer: And also librarians have a broad knowledge of the university system itself, how the university functions, and which departments interact well; that is also considerably beneficial for collaborations and partnering.

There are also issues of archiving the things that are created—not just the finished projects themselves but the artifacts that are made along the way and the thinking that occurs as the projects move forward. One of the great tragedies of digital media is that we do not have inventors' letters like in the past. There were all sorts of people that have curated from innovators, using artifacts and notebooks, etcetera, and now that we are innovating electronically, information is easily lost as computers change or e-mail subscriptions run out and are replaced with new accounts. Librarians are asking very important questions about that, such as how to innovate in a way where not just the artifact is kept but also the knowledge about how you got there.

José Zagal: And that causes innovators to ask questions about archiving knowledge that we otherwise would not have thought to ask ourselves.

Roger Altizer: Agreed.

José Zagal: If everything were to blow up tomorrow, what would you want to have stay? We often focus on the games themselves or the papers that we write about them, but everything in between can be relevant. It is the librarians that are asking the questions about how we can dig deep and capture that knowledge. In the context of *edu-tainment*, one saw year after year the same developments that had been done prior. Clearly, there was a problem that either the right papers were not being published or the right knowledge about earlier attempts was not available. And if, for example, you have an idea about building on a game that was developed in 1984, it would be extremely difficult to find any informative details about its development.

Roger Altizer: It is ironic and exciting that the stereotype of librarians is that of gatekeepers of information and that information should look like books, but that is no longer reality. Challenges of how to archive and discover developed software are questions that academics and people in the industry are asking ourselves and that librarians are in the forefront of determining. In our collaboration at The GApp Lab, we have the great privilege of working with an Innovation Librarian, so there is actually a librarian who is dedicated to these types of questions. I am coteaching a class with her; she is working with our students to write articles and posters on these topics, she has helped us find information, and she has also served as a liaison to the EHSL at large for us. That has made life incredibly easy for us. It is very easy to get people excited about working on projects, it is much harder to get hours of people's time, so to have someone who is dedicated to it has made things a lot easier.

José Zagal: It makes me wish that every research group could have a dedicated librarian.

Roger Altizer: Yes, absolutely, that would be brilliant.

Interviewers: Where do you see The GApp Lab going, in theory, and then in actual physical space?

Roger Altizer: In the next several years we expect to see the relationship between The GApp Lab and its various collaborating units become more formal. Right now it is similar to a start-up, and the success of a start-up depends on having resources available and people with lots of energy to take advantage of those resources. That has been the key to our success. The popularity of the lab has been enormous, and we very quickly found ourselves reaching capacity. Some of the questions moving forward are how to prioritize new projects and use of resources and what are the best practices in terms of developing and archiving projects.

Another is how to create a process where you can easily repeat certain aspects of the work rather than reinventing the wheel for each project. There are many complicated technical and administrative questions this applies to, such as how to create a test app that is HIPAA- compliant and can run on U of U servers. That requires a considerable amount of coordination of a lot of people as well as a level of understanding of the needs of the system and the laws. Rather than reinventing that on every project, we need to make it into an easily repeatable process.

Also, The GApp Lab has uncovered a massive need in the Health Sciences, which is that there is a large number of people who want to do app projects for entrepreneurial, clinical, or research reasons, and there needs to be systems to support them. Right now they are turning to The GApp Lab for this. That is another great reason to be in the EHSL as well since people often turn to it for help with information needs. It is an opportunity for libraries to expand their mission to collaborate with people or create labs to do this kind of work for health science systems. We live in an era where people want to create software, and we need to create methods that allow them to do that and support them in doing so. That is what the next several years will look like—figuring out how to meet the needs of students and faculty while making sure that research needs and the development needs of the health care system are met.

Interviewer: Do you see the space becoming physically different?

Roger Altizer: Yes, The GApp Lab is slated to grow both in footprint and in the number of students and staff. What would a similar lab need, if someone were to ask that question . . . it would need a space to develop ideas where students can sit, such as traditional computer rooms, and also war rooms with whiteboards and markers to hold consultations, experimental spaces to build hardware, and offices to accommodate those who will be doing the actual science work and writing but also need proximity to those involved in the process. Locate that in the middle of a place that is safe and supportive, and you have a winning lab setup, which is exactly what we have done. All sorts of great energies occur.

José Zagal: A challenge for the future is to see if The GApp Lab can become a new model for software development in an academic setting. For instance, instead of semester-by-semester hiring of students to work on a specific project, could The GApp Lab be more closely integrated with university programs to offer a residency in a development lab on an annual basis for graduate students in education, medicine, and engineering?

Roger Altizer: That is a model we are actively pursuing to enable students to be here longer and ask more serious questions with more challenging projects.

Interviewers: What do students think about being located in a library?

Roger Altizer: They think it is funny to be located in a library. The initial concern about having to be quiet has dissipated, and it is working out great. Students are now encountering the realities of being part of a managed system with rules and relationships. It is giving them an experience that closely resembles the work environment and is teaching them to be good neighbors in a safe place where everyone is nice and supportive. It helps them work through those kinds of challenges.

José Zagal: They do not really think of it as a library, which speaks to the nontraditional character of the EHSL space. And I think it is generational; most of the students are in their early twenties and therefore do not have a set idea of what a library is or should be.

Interviewers: Would you recommend other gaming programs to be located in a library? If not, why not? And, if so, why, and what recommendations do you have for others?

Roger Altizer: I think a lab like this within a library works out great as long as the library and the people that work within it want the collaboration. That seems like a perfect situation. We have had labs all over campus, for teaching and other things, and sometimes it is a good fit and sometimes it is not, and that tends to be due to both the space itself and the people that work within it.

José Zagal: It seemed when The GApp Lab was created, the EHSL was going through the process of reinventing its physical space, so the timing was perfect for testing a new collaboration like this.

Roger Altizer: Yes, it seems a lot of libraries are asking themselves what the future looks like for them, and since information is changing so rapidly, it allows for being involved in the creation of that future. In this case, it has been a really good fit, obviously, since we have had such great success.

Interviewers: Do you ever see a blend profession of a librarian (information specialist) and gaming? What would be the advantages of this mixed skill set?

Roger Altizer: There are a lot of libraries that are asking questions about games and software and particularly the archiving aspect of games since the relationship to entertainment poses more complicated questions about preservation. Questions are being asked, such as, "What do games look like in a library?," "What is the artifact that you keep?," "How do you get people access to it?," "Does it look like a play lab that people visit?," and "Do you keep the games with all the original hardware, meaning you have to track down old cathode ray tubes etcetera and try to re-create the original environment, or can you emulate it all?" These are the same kinds of questions that are being asked about things like film: do you keep the original artifact so that people check out canisters of actual film that then has to be viewable on appropriate hardware, or do you digitize and keep them in an archive, and what complications arise from that? With gaming as with other aspects of entertainment information, there are a whole myriad of archiving questions that need to be considered.

Add to the archiving complications the fact that medical games and apps are rapidly growing fields—games as a general field is growing massively—medical games and apps are growing at an uncontrollable rate. Having a subject specialist that could answer or find the answers to questions such as whether there has been a similar game created or attempted in the past and what were the particulars of that past attempt would be incredibly valuable. Right now we could get grant funding for a particular project and find it extremely difficult to find prior projects in the same field. There is currently no method in place to even start a search like that, no best practices for a student on how to find that kind of information. They would turn to Google or word of mouth. An information specialist would have means for finding systematic reviews and information about a game that has been built like the one you are currently building, and that would be an extremely valuable person.

José Zagal: A media specialist librarian can find the answer to questions such as which television shows during a certain decade had all-male casts. We need a similar someone for gaming. There are questions outside the actual list of games covering a certain issue; there is also the idea of which ones actually worked, which worked best and why, what about those that did not get launched, and the whole question of documented failures. You would want to know which games worked well in the past so you could analyze why, look at the details of their content, and decide what you would want to emulate in new games being created.

Roger Altizer: There is a huge amount of room for research and service for a librarian that specializes in games.

Interviewers: What have we not asked you that you want to have included in this chapter that will be read by librarians?

José Zagal: Looking at how The GApp Lab might have failed is important to consider. To be successful, these kinds of innovation spaces need to have specific things that need to happen in order to maintain a high energy level. The GApp Lab is a lab with specific deliverables that have to get done by certain dates. The size of the projects should be considered. Numerous small projects stimulate enthusiasm since something is always working well, and therefore there is always something interesting going on. Multiple smaller projects allow for more collisions, and there is always at least one project at an exciting stage. A single large grant on a single project does not stimulate as many small successes that keep enthusiasm high, particularly if that project ends up not working and is unsuccessful. That is why having one huge $5 million grant for a single project does not work nearly as well. You need multiple projects with diversity to engage the largest group of people.

Roger Altizer: I totally agree; having five to nine projects a semester offers diversity and keeps students engaged. It allows you to get wins on a fairly regular basis, which keeps energy levels high. Having many things going on allows you to get to best practices, and doing something well is always more fun and more exciting.

Interviewers: Thank you both for your time and for the insightful comments about how The GApp Lab has progressed and benefited from its location within the EHSL.

Chapter Nine

Information Needs of Medical Digital Therapeutics Personnel

Tallie Casucci

Apps and games have risen in popularity within the past 10 years. Originally, mobile apps were used to improve one's general productivity and communications. With development tools and mobile devices becoming ubiquitous, the utility of apps and games has expanded. Industries such as health care are leveraging the prevalence of mobile devices and apps and the entertainment and education qualities of games. By capitalizing on these attributes, health care professionals are engaging patients to track health care indicators and outcomes, encouraging better health behaviors and wellness activities, and using them to train and develop skills.

Health sciences librarians have the skills and expertise to collaborate with developers of these medical games and apps, or *medical digital therapeutics* (MDT). Such specialty librarians ensure that evidence-based information is included and documented in the MDT development process. This chapter provides a case study of an embedded librarian or an innovation librarian (IL - innovation informationist) and her contributions and impact on the University of Utah's (U of U's) MDT units.

SETTING

Spencer S. Eccles Health Sciences Library

The Spencer S. Eccles Health Sciences Library (EHSL) is one of three libraries at the U of U. EHSL serves the U of U Health Sciences, which includes the Colleges of Health, Nursing, and Pharmacy; the Schools of Dentistry and Medicine; and the U of U Health Care's hospital, clinics, and several centers and institutes. EHSL's mission is "to advance and transform education, research, and health care through dynamic technologies, evidence application, and collaborative partnerships. The library contributes to the success of health professionals, students, researchers, and the community" (http://library.med.utah.edu/about).

MDT at the U of U

In January 2014, EHSL formed two partnerships with MDT developers. The first, The GApp Lab, is a graduate student game development lab, and the second, the Games4Health Challenge, is a global student game development competition. The GApp Lab students are part of the U of U's Entertainment Arts and Engineering (EAE) Program, which is nationally ranked number one in graduate and number two in undergraduate game studies. Both The GApp Lab and the Games4Health Challenge have grown significantly in recent years. The GApp Lab averages 15 to 20 completed projects a year, and the Games4Health Challenge has greatly increased student participation annually.

The GApp Lab, formally the *Therapeutic Games and Apps Lab*, is a partnership between EHSL, EAE, and the Center for Medical Innovation (CMI). The CMI and The GApp Lab are housed in the Synapse on EHSL's Garden Level, creating an epicenter for medical innovation (chapter 3). The GApp Lab connects game developers to medical and health professionals on collaborative projects that are changing the future of health care. The GApp Lab partners on projects with U of U affiliates, community organizations, and industry entities. These project collaborators provide funding for the students and background information for the projects' research fields. Each semester, The GApp Lab normally trains 25 to 35 students who work as teams on approximately 6 to 10 projects. Each project team is comprised of an interdisciplinary group of students including an artist, engineer, producer, and quarter-time research associate. The GApp Lab is managed by staff and directed by one EAE faculty lab director. Other EAE faculty members assist on specific projects and student teams (chapters 8 and 12).

The second MDT program, the Games4Health Challenge, originates from the Sorenson Center for Discovery and Innovation within the U of U David Eccles School of Business. This competition occurs during the spring semester of the academic calendar and is open to any university student team in the world. Teams are tasked with developing a game related to health or medicine, either beta versions or simple wire-frame mock-ups.

As medical device and MDT development expanded within the U of U Health Sciences, EHSL recognized a potential need for focused innovation information support and partnership. The EHSL director reallocated funds for a new position to test the concept. The innovation and research associate was a one-year position to determine the necessity of focused innovation involvement. It quickly became apparent that this was a viable position for EHSL, and a permanent tenure-line IL faculty position was created (chapter 11). A national search ensued with the former innovation and research associate being hired as IL starting in January 2015.

The IL collaborates with health sciences innovators to discover evidence, competitive intelligence, market research, and other information to further develop ideas and commercialize products. The IL works directly with faculty and students from CMI, The GApp Lab, the Games4Health Challenge, and other innovation groups (chapter 11). These partnerships evolve each semester on varying levels. This chapter highlights the information needs of MDT developers and how the IL has addressed those needs.

INFORMATION NEEDS OF MDT DEVELOPERS

Game developers face several challenges to find relevant information. The scholarly literature of gaming research is published across multiple disciplines. The gaming research literature covers all aspects of games, such as game design, psychological impact, educational outcomes, historical perspectives, humanities influences, socioeconomic factors, and health effects. Due to the multidisciplinary nature of gaming research, it takes time to find relevant literature for game scholars and game studies students. This provides an excellent opportunity for information professionals, who can easily navigate the expanse of citation databases, gray literature, and other pertinent information resources, to assist MDT development.

Another challenge for game developers is discovering prior games on similar topics. Many games are lost as technology advances, platforms and software become obsolete, and game developers stop development or support of older games. As game developers investigate prior games, they often find editorial opinions, brief game reviews, and incomplete snippets of recorded game play. This makes it difficult to learn and build on past experiences. Librarians are very interested in creating game repositories to address this archival issue; however, it is an incredibly difficult undertaking with storage and technology constraints.[1]

Additional challenges are encountered when medical and health games are the focus. MDT developers have to be concerned with the Health Insurance Portability and Accountability Act, patient privacy, Food and Drug Administration approval, clinical data management, and much more. Many medical professionals want evidence-based approaches for proving MDT impact and efficacy before recommending MDT to patients. This adds another layer of complexity for MDT developers. They must not only create an alpha game but also demonstrate clinical validity. Most game developers are not familiar with these topics, creating an opportunity for health sciences librarians to teach these aspects of MDT development. With their network of contacts and involvement on other projects, health sciences librarians can refer MDT developers to relevant information resources or connect them to area specialists within their institutions.

Although libraries and librarians cannot solve all of these challenges, they can alleviate some information gaps for MDT developers. The IL discovers MDT developers' information needs either through direct queries or by simply listening to their complaints or challenges and offering possible solutions. The next few pages outline how the IL has helped The GApp Lab and Games4Health Challenge members to meet their information needs.

The GApp Lab's Information Needs

The IL is The GApp Lab's primary librarian for anything related to the EHSL or information. At the beginning of each semester, the IL facilitates access to the EHSL building and its e-mail listservs and creates photo directories. One photo directory is of EHSL employees, and the other directory is of The GApp Lab students. These directories allow EHSL and The GApp Lab students to recognize and talk to each other in the hallways. The IL introduces herself during The GApp Lab orientations, which may include a tour of the EHSL. The GApp Lab students ask the IL for information regarding their games during both weekly meetings and personal consultations. Since the IL's office is nearby, students generally stop by without appointments. The type of information and involvement needed depends on the project.

In the spring of 2014, The GApp Lab created *Open EHR App*, an open-source code electronic health record (EHR) app designed for developing countries (http://gapp.eaemgs.utah.edu/wordpress/ehr). The *Open EHR App* enables patients to record and maintain their medical health histories. Since patients own their personal health histories, they do not need paperwork transferred between various health professionals and clinics. Patients in developing countries usually see different health care providers, so the *Open EHR App* empowers patients and streamlines the medical tracking and recording processes. Additionally, health care professionals can add notes directly into the app for both patients and other health care providers to reference.

Prior to this project, the EAE students had not seen the front or the back end of an EHR and therefore wanted to see examples of EHRs. Under the IL's guidance, a part-time EHSL Public Services staff member located EHR pictures and lists of EHR components. Additionally, the IL connected the students to the U of U Health Care's EHR developers.

The students also requested information about patient privacy laws and recommendations for developing countries. For this information component, the IL spent more than 26 hours compiling and gathering literature. The IL investigated patient privacy recommendations from various organizations. Additionally, she provided national, global, and country-specific regulations and laws.

For another project, The GApp Lab created a game, *Save Your Bacon!*, designed to encourage college-age individuals to purchase health insurance. Players choose different activities and attempt to make it through a month without becoming injured. Eventually, the players become injured, and the game informs them of the associated injury costs, with and without health insurance. The IL provided the health and insurance data for this game, which were gathered through 12 hours of mediated searching. The data included statistics on the types and rates of occurrence for nonfatal injuries and the average health care costs for injuries both in the United States and in the state of Utah.

During the first two semesters, a private LibGuide was used to share information between the IL and The GApp Lab students. Currently, information is shared via e-mail or Google Drive. The EAE Program heavily utilizes Google products for organizing and tracking projects. The IL was added to The GApp Lab's Google Drive folders and can share links and citations directly with the students. The above examples illustrate more involved information requests; most requests are not as time intensive.

Starting in the fall semester of 2014, The GApp Lab added quarter-time student research assistants (RAs) to each project team. An RA's role is to write and submit abstracts and papers about The GApp Lab projects to professional conferences and journals. Initially, the IL, EHSL director, and lab director met with the RAs on a weekly basis at the beginning of the semester to review progress and provide insights. A list of potential conference outlets was cocreated by the RAs, lab director, and EHSL faculty. Additionally, other EHSL faculty have presented and led discussions on professional writing topics. For example, many RAs were not familiar with the components of conference abstracts, and EHSL faculty helped to guide how to organize abstracts. This RA component of The GApp Lab remains in development. A possible future direction is converting the RA program into a graduate writing class on serious games.

While copresenting[2] with the lab director at an annual gaming research conference, the Digital Games Research Association (DiGRA), the EHSL director and the IL discovered that many game developers want more information about games and their development processes. As mentioned earlier, game researchers desire in-depth information about the game development teams. During an interactive DiGRA session,[3] game researchers recognized that some games receive funding for multiplayer game versions, but the source of funding is not easily discoverable. Thus, potential funder biases, if they exist, are not easily identified. Additionally, other concerns about the development process, funding, and authorship remain a mystery for game researchers, especially for games developed by companies. By tracking the intellectual processes, funding sources, and contributing roles, researchers and others can better evaluate the validity and potential bias of a game.

This knowledge gap is being addressed through two means. The first is e-channel (chapter 13), a multimedia platform of innovation-related content. The GApp Lab's *wrap kits*, which include documents such as team members, project time lines, design documents, artwork, videos, and source code, are contained on e-channel. The second is through annotated bibliographies. The GApp Lab and other serious MDT projects are encouraged by the IL and lab director to develop such documentation. With an annotated bibliography, others can learn about factors influencing MDT development.

In the spring of 2016, the IL and lab director cotaught a special topics class on serious games, focusing on virtual reality (VR) (chapter 11). For the midterm presentation, students pitched their VR game ideas for their final projects and produced a one-page description and design concept for each VR game, and every team created an annotated bibliography. These annotated bibliographies provide written documentation about the research behind the VR game, helping future developers, researchers, and players understand the development process and background evidence. Moreover, the students learned and illustrated how their VR games contribute to and fit within the current body of knowledge. Some students met with the IL to discover literature on their topics. For many students, it was their first time identifying primary literature on medical, psychological, or educational topics.

Another information need that The GApp Lab members and their project collaborators face is obtaining funding. The MDT projects need to move from idea to a beta game and eventually to an alpha game. Obtaining funding for these sequential steps can be challenging. The lab director and staff have included the IL in these project collaborator and future project collaborator meetings. The IL offers expertise in using information databases to locate funding opportunities for projects. Additionally, the IL guides the researchers to think more broadly about their game ideas to find additional funding opportunities. For example, terms such as *decision support tool* may resonate with traditional grant funders more than an MDT.

The GApp Lab solicits EHSL patrons for *game testing*. By offering game testing in a library, the game developers have access to more game testers with varying gaming skills. The GApp Lab's *Reflex Speed* is one game that EHSL's patrons and employees tested several times. *Reflex Speed* is a game designed for athletes and coaches to enhance reflex abilities and track overall performance. The game utilizes a quadrant to indicate the side of the body (left or right) and body part (hand or foot) that should react. For example, if an image appears in the upper left quadrant, the player should push a button on the left-hand controller.

In addition to the directories and game testing opportunities, another way the IL encourages interaction between other EHSL employees and The GApp Lab are presentations. Once a year, The GApp Lab students present at the EHSL all-staff meeting. They provide a brief overview of the completed projects and sometimes show a video of the game play. These presentations provide The GApp Lab students an opportunity to talk about their work before a large audience and receive feedback before meeting with project collaborators in final project handoff meetings.

Games4Health Challenge's Information Needs

The Games4Health Challenge began in the spring of 2014 and continues to grow annually with 24 teams in 2015 from three countries and 121 teams in 2016 representing 12 countries. As stated earlier, the teams create either a beta version or a wire frame of a health game. Games4Health Challenge issues *sponsor challenges* for teams; these serve as possible themes. For example, an insurance company sponsored a challenge for games to discourage texting while driving. Five teams designed games for this sponsor challenge and competed against one another for this sponsor's monetary award. The Games4Health Challenge is organized by two student leaders and a faculty director.

Each competition year requests different deliverables from competitors. In 2014–2015, teams submitted a game or wire frame, a business plan, a clinical trial proposal to test the efficacy of the game, a teaser video, and a *team details* form. In 2015, the Games4Health leaders picked specific software for creating wire frames; however, in other years, teams could choose their own wire-frame software platform. In 2016, the teams were required to submit two videos to compete in the Games4Health Challenge. The first video was a teaser video that included game play and a review of the significance and purpose of the game. The second video was the commercialization video, which included components of the team's business plan and clinical trial proposal (if applicable).

To support student competitors, the IL created a *Games4Health Challenge* LibGuide[4] that is updated annually with relevant information resources, reflecting the competition requirements and deliverables. In addition to medical game research databases, the LibGuide includes business plan and clinical trial information. Many of the student competitors' educational backgrounds are in business and game studies, so the clinical aspects and even health literacy components of health games are often unknown. By providing the teams with this information, they quickly learn about these topics and can apply them to their games.

For the inaugural 2014 competition year, the Games4Health Challenge student leaders and the faculty director requested a matching site,[5] a way to help teams find potential team members and match teams to industry and academic mentors. The IL created this site to function similarly to a standard *dating site* using the U of U's e-commerce platform. Anyone could *purchase* mentors or post ads for needed team members. This matchmaking site was poorly advertised and utilized compared to the effort needed to maintain the back end, so the platform was discontinued in 2016. In retrospect, after creating the platform, the IL should have given administration rights to the student leaders. This may have encouraged the student leaders to advertise the service to the competition participants. Furthermore, another platform, such as a discussion forum, may have been an easier matchmaking solution for team formations. For the team-to-mentor matching component, a simple request system may be more effective. The IL utilized a simple request system for another competition that worked well.[6]

The Games4Health Challenge student leaders asked the IL for advice on ways to reach U of U Health Sciences students. Many of the existing U of U student team members were from the School of Business or EAE Program. The IL recommended several ways to advertise and recruit student teams from the U of U Health Sciences. With the help of part-time EHSL Public Services staff, flyers were hung in the Health Sciences buildings. Additionally, a Games4Health Challenge announcement slide was included in the EHSL advertisement display slides that rotate on several display monitors in the EHSL.

The IL and EHSL director meet with the Games4Health Challenge director semiannually to discuss partnership opportunities. After the inaugural 2014 competition, the Games4Health Challenge director really wanted to expand the competition to other university student teams, not only those affiliated with the U of U. The IL compiled a list of journals whose editors may be interested in competitions for developing health games. With this list and personal contacts, the director pursued focused outreach. Again, in preparation for the 2016 Games4Health Challenge, the director requested information about university gaming labs, especially those working on medical or health projects. The IL gathered a list of university gaming labs to contact for potential collaboration. These competitive intelligence reports helped the director expand the competition's reach and influence.

EHSL has also partnered with the Games4Health Challenge leadership to collect team materials for e-channel (chapter 13). Currently, the Games4Health Challenge collection consists of videos; however, if future years require written materials (e.g., team details, business plans, clinical trial design, etc.), e-channel will also host such documentation.

In 2014, the Games4Health Challenge cohosted two events with the U of U Entrepreneur Club in the EHSL's Synapse (chapter 3). The first event was a two-day start-up challenge called *Jump Start Innovation*. The first evening, students self-divided into teams based on a problem. On the second night, the teams pitched their ideas to solve these problems. The second event, *FrankenApp*, allowed students to mix and match different app features to re-create a better app.[7] For both events, students competed for cash prizes to further develop their ideas. During the events, the IL welcomed the students to the Synapse and invited them to utilize the space to work on collaboration projects.

Other MDT Projects

With the success of these two MDT programs, others at the U of U have voiced interest in developing their own MDTs. The IL created a LibGuide, called *Game Development and Research*,[8] for serious game development and research in the U of U Health Sciences. The guide highlights literature databases, development tools, U of U MDT development entities, and U of U MDT examples. It is a starting place for those new to serious games and apps within the health sciences context.

Another information need of MDT developers is tools necessary to vet medical and health games and apps. EHSL's faculty and staff are involved with two projects that offer such vetting components. The EHSL faculty organize and host an event, Appy Hour, which highlights a featured app.[9] The presenter simply navigates through the app and answers audience questions. Appy Hour is an informal time for faculty, staff, and students to network, enjoy refreshments, and learn about apps.[10] The second project is the U-Bar, an app bar in the U of U Health Care's consumer health library operated by EHSL. Patients and family members can see, test, and receive help downloading and using apps on their personal devices (e.g., phones and tablets) and wearable technologies. Training for clinicians on the U-Bar apps is being offered. By teaching the clinicians about the apps, they are encouraged to refer patients to the U-Bar. With Appy Hour and the U-Bar, several faculty and clinicians have questioned the process for how apps are selected for inclusion. Their concerns of currency, accuracy, and relevance are trickling down to the MDT developers to be incorporated into locally created apps. As MDT developers investigate prior apps and games, they need to think critically about how their projects add to established research.

SUMMARY

Health sciences librarians can be a significant partner to MDT developers, as they have a great depth of knowledge concerning the health sciences context and can easily direct, explain, and teach principles or practices to MDT developers. The majority of MDT developers lack experience with the health sciences, health care, and research-based fields, so many struggle with concepts familiar to experts within these areas. As a neutral and service entity, health sciences librarians not only provide background knowledge but also can direct MDT developers to resources relevant to their projects, such as prior research, experts and mentors within a given field, and other key services and tools.

NOTES

1. Eric Kaltman, Noah Wardrip-Fruin, Henry Lowood, and Christy Caldwell, "A Unified Approach to Preserving Cultural Software Objects and Their Development Histories: A Case Study in Academic Computer Games," 2014, Center for Games and Playable Media, University of California, Santa Cruz, http://escholarship.org/uc/item/0wg4w6b9 (accessed May 20, 2016).

2. Tallie Casucci, Jean P. Shipman, Roger A. Altizer, and John T. Langell, "Revolutionizing Game Creation Partners: Health Care Professionals, Including Librarians, and Game Scholars Unite" (paper presented at the Digital Games Research Association Annual Conference, Snowbird, UT, August 4, 2014).

3. Tallie Casucci, Jean P. Shipman, and Roger A. Altizer, "Unravelling Medical Game Research: Informing Players of Foundational Evidence" (paper presented at the Digital Games Research Association Annual Conference, Snowbird, UT, August 5, 2014).

4. Tallie Casucci, "Games4Health Challenge: Welcome," 2016, http://campusguides.lib.utah.edu/games4health (accessed February 12, 2016).

5. Tallie Casucci, "The Art of Matchmaking: Connecting Teams to the Right Mentor" (paper presented at the VentureWell Open 2016 Conference, Portland, OR, March 4, 2016).

6. Ibid.

7. Tallie Casucci, "FrankenApp—Tuesday, October 21, 5–8pm in Room 25," *Eccles Library Blog*, October 20, 2014, http://library.med.utah.edu/blog/eccles/2014/10/20/frankenapp (accessed February 12, 2016).

8. Tallie Casucci, "Game Development and Research: Welcome," 2016, http://campusguides.lib.utah.edu/games (accessed March 25, 2016).

9. Spencer S. Eccles Health Sciences Library, University of Utah, "Mobile Devices—Tablets, eReaders and Apps: Appy Hour," 2016, http://campusguides.lib.utah.edu/EcclesMobileDevices/appyhour (accessed March 25, 2016).

10. Tallie Casucci, Joan M. Gregory, and Jean P. Shipman, "Appy Hour: Health Sciences Professionals Learn about Apps," *Medical Reference Services Quarterly* 35, no. 3 (2016): 251–58.

Chapter Ten

Medical Innovation Competition Information Support

Erin Wimmer, Tallie Casucci, Jacob Reed, Nathaniel Rhodes, Benjamin Fogg, Thomas J. Ferrill, David Morrison, Alfred Mowdood, Darell Schmick, Mohammad Mirfakhrai, and Peter Jones

There are three libraries at the University of Utah (U of U): the J. Willard Marriott Library (Marriott), the Spencer S. Eccles Health Sciences Library (EHSL), and the James E. Faust Law Library (Faust). Although these are separate entities, the libraries collaborate to avoid duplication of effort and to create more meaningful projects, events, and initiatives together.

The Marriott, dedicated in 1968, is "the flagship academic library for the Utah State System of Higher Education and the largest state-funded academic library in the six-state region."[1] The Marriott is located in the center of the Main Campus with nearly 1.5 million annual visits. Its faculty and staff have expertise in science, engineering, business, social sciences, fine arts, and the humanities. The Knowledge Commons staff, located within the Marriott, provides research and technology assistance and expertise (http://www.lib.utah.edu/services/knowledge-commons).

The EHSL partners with the Health Sciences, which includes the Colleges of Health, Nursing, and Pharmacy; the Schools of Dentistry and Medicine; the U of U Health Care Hospital and Clinics, and many centers and institutes. The mission of EHSL is to "advance and transform education, research, and health care through dynamic technologies, evidence application, and collaborative partnerships. The library contributes to the success of health professionals, students, researchers, and the community."[2] EHSL hosts the Regional Medical Library for the MidContinental Region of the National Network of Libraries of Medicine and the National Library of Medicine's National

Training Office. On EHSL's Garden Level (the lower level) is a technology-rich collaboration space with librarian services and the Center for Medical Innovation (CMI), Therapeutic Games and Apps Laboratory (The GApp Lab), a Skills Center, and the Gary L. Crocker Innovation and Design Lab (chapters 3, 6, 8, 9, and 12).

The Faust serves faculty, staff, and students of the S. J. Quinney College of Law. The Faust is dedicated to the teaching, research, and service mission of the College of Law. As the legal information repository for the U of U and the largest public law library in the state, the library also serves the legal information needs of the broader community. Librarians work closely with other law faculty, and the librarians with JD degrees teach legal research classes.

One specific U of U program that requires the skills and expertise of all three libraries is the Bench to Bedside (B-2-B) competition. B-2-B is an annual medical device competition for U of U students. This competition initially started by bringing medicine, business, and engineering students together to discover solutions for unmet clinical needs. Now, B-2-B includes students from many disciplines. With $500 of prototyping funds, B-2-B interprofessional team members develop a medical device that is judged at a final competition for monetary awards. The award monies are used by B-2-B teams to further prototype and bring their devices to market.

Even though the U of U has a long history of medical innovation and entrepreneurship (table 10.1), the libraries had never formed a unified team to reach this audience. In 2012, the libraries created a dedicated team, the Libraries Innovation Team (LIT), to promote the libraries' resources and services to innovators. LIT includes members from all three U of U libraries and inspires, encourages, and supports innovation and related initiatives. This chapter illustrates LIT's role with innovation-specific outreach, resources, and services, specifically for the B-2-B competition.

LIT

LIT's mission is three-pronged: promote campus innovation, provide resources and tools, and conduct outreach to innovative U of U partners. LIT has used formal meeting times to develop the mission and to share information about new and specialized services at each of the libraries that may benefit innovation groups. For example, LIT provides tours of library innovation spaces and demonstrates innovation-related databases and tools. Since LIT members span three libraries that are geographically separated, a LIT e-mail listserv was created to share information between meetings and serves as a central contact method for B-2-B teams and other innovators.

Table 10.1. University of Utah Medical Innovation History.

Year	Innovation at the University of Utah (U of U)
1936	The S. J. Quinney Law Library becomes its own entity. The law collection begins in 1905, with the first part-time law librarian hired in 1927.
1967	Utah artificial organs program begins.
1968	David C. Evans and Ivan Sutherland, professors in the Department of Computer Science, found Evans & Sutherland. Evans & Sutherland pioneer the computer graphics field that included simulation. J. Willard Marriott Library (Marriott Library) building opens to the public. The technology transfer office is created and eventually renamed Technology and Venture Commercialization.
1971	Spencer S. Eccles Health Sciences Library (EHSL) building opens, and new medical librarians are hired (before this time, medical librarians worked within the School of Medicine and Hospital).
1972	MEDLARS terminal is installed in EHSL. Homer Warner, MD (class of 1949); Reed Gardner, PhD (class of 1968); and Alan Pryor (class of 1972) introduce HELP, the world's first clinical decision support system.
1979	A surgical team led by Clifford Snyder, MD, and Theodore Roberts, MD, performs the first successful separation of brain-conjoined twins. Dr. Snyder provides funds for an annual EHSL lecture. The Institute of Biomedical Engineering, led by Dr. Willem Kolff, MD, PhD, unveils the wearable artificial kidney, which could literally be worn around the waist, allowing dialysis to be performed at home.
1984	Marriott Library is designated as a Patent Depository Library.
1985	Ted Stanley, MD, pioneers human use of the drug fentanyl and receives Food and Drug Administration approval to administer it in the form of a lollipop. John Dixon, MD, and Richard Straight, PhD, organize the Laser Institute, one of the first laser surgery labs in the country. The first Mac computer on campus is installed in EHSL.
1986	Suzanne Stensaas, PhD, begins using one of the earliest forms of computer-based education at EHSL. *Slice of Life* videodiscs introduce a brand-new way of teaching and sharing medical images. It also ends the need to carry around carousels full of Kodachrome heavy glass lantern slides.
1990	EHSL's InfoFair begins. As an annual event, InfoFair provides up-to-date information on computer services, applications, and resources in the health sciences.
1994	Geneticists Mark Skolnick, PhD, and Lisa Cannon-Albright, PhD (class of 1988), lead a U of U research team to successfully clone the BRCA1 breast cancer gene.
2001	Lassonde Entrepreneur Institute is established with funding from Pierre Lassonde.

Year	Innovation at the University of Utah (U of U)
2007	Entrepreneurial Faculty Scholars are formed. Mario Capecchi, PhD, Distinguished Professor of Human Genetics, is awarded the Nobel Prize in Physiology or Medicine for pioneering gene targeting in mice.
2010	Association of University Technology Managers recognizes the U of U for fostering research-based inventions into new start-up companies. Bench to Bedside competition is started by the Center for Medical Innovation; holds first competition in 2011.
2012	Center for Law and Biomedical Sciences is created to support the U of U's role in biotechnology, intellectual property, and the patent disclosure process. Libraries Innovation Team is formed.
2015	The S. J. Quinney Law Library becomes the James E. Faust Law Library and moves into the new College of Law building.
2016	Lassonde Studios opens, an innovative dormitory with innovation spaces for more than 400 students.

Promote Campus Innovation

An *Innovation Guide*[3] was created to promote resources and services available to innovation groups. The *Innovation Guide* provides information on concept generation, locating innovation and creativity spaces at the U of U, and promoting innovative resources. It also identifies centers, programs, and competitions with which innovators may engage. This *Innovation Guide* receives a significant number of views throughout the year, not only during competitions.

LIT also promotes innovation by highlighting competition events and opportunities through social media, blog posts, and on display panels in the libraries. For example, in 2016, the EHSL created a blog post to recruit judges for an international medical game competition.[4] Additionally, the EHSL and Marriott advertised the kickoff event for a health video competition through their rotating display panels and social media platforms. Finally, after each B-2-B competition, a LIT member writes a recap of the event that is used in the libraries' various newsletters to promote LIT.[5]

Provide Resources and Tools

Early in LIT's involvement with the B-2-B program, the student leadership expressed a need for a better system to provide information, communicate announcements and alerts to teams, and facilitate team coordination. A LIT member suggested using Canvas, the U of U's learning management system, based on previous successes applying this tool for team collaboration. B-2-B leadership was interested in pursuing this idea, so the LIT member created a prototype based on a packet of information provided by the B-2-B student leaders. After some feedback, the Canvas course was implemented and became the primary communication venue for B-2-B teams. B-2-B leadership and LIT made announcements through Canvas that were distributed to all of the teams. Each team also had its own work space within the Canvas course for communicating and sharing relevant resources. Finally, a *team formation corkboard* allowed interested students to post ideas and find other potential team members online rather than at in-person events. LIT maintains a presence in the B-2-B Canvas course, including LIT's e-mail, the *Innovation Guide* link, and other library resources relevant to B-2-B.

Innovators commonly use a tool called the *Business Model Canvas*[6] to generate ideas focused on the key steps related to a new product. These steps include value propositions, customer segments, key partners, key activities, key resources, customer relationships, channels, cost structure, and revenue streams. The *Innovation Guide* and e-channel (http://library.med.utah.edu/e-channel/innovation-vault-business-model-innovation) provide additional information on this model[7] as well as suggestions for databases to use when considering each of the steps. For example, in the Cost Structure step, B-2-B teams are encouraged to use the *Material ConneXion* database to browse more than 7,000 materials and design processes used in a variety of fields. They are also directed to *ThomasNet.com*, a free website for identifying suppliers, and *MarketResearch.com Academic* and *IBISWorld*, which include coverage of market trends, industry reviews, market share, competitive analysis, and more.

To further facilitate B-2-B and other innovation groups in working with the *Business Model Canvas*, EHSL printed several large, reusable vinyl canvases of the model. Some of the canvases are hung in collaboration spaces used by B-2-B groups, while others circulate. Using the canvases, students collaborate as teams to build a cohesive business model.

EHSL and Marriott have developed services and spaces for innovators to prototype their products. A variety of 3D printers and scanners, a materials collection, and production tools are available for innovators to create prototypes. These spaces are the result of collaborations between the libraries and other entities invested in innovation. For example, the Gary L. Crocker Innovation and Design Lab is a joint venture between the CMI and EHSL.

The popularity of these tools and services has prompted the need to expand innovation spaces in the libraries. From the initial purchase of a 3D scanner at the Marriott to a full suite of scanners, printers, Pantone color products, and material types, Marriott continues to develop the tools and services it offers. The Gary L. Crocker Innovation and Design Lab at EHSL originated with a 3D body scanner, donated by EHSL, and 3D printing tools[8] and has grown to include additional prototyping tools and resources, such as International Organization for Standardization standards (chapter 6).[9]

B-2-B teams frequently use these prototyping tools to produce device models. One B-2-B team member used the 3D printers at the Marriott to print a hose adaptor for his team's prototype. At the B-2-B competition, LIT members observed the prototype in action, with the 3D-printed model alongside several others, including one fabricated from steel. Often, these low-cost printing solutions are used as proofs of concept to ensure that the model is satisfactory, before being sent off for higher-end, more expensive fabrication production.

Because these tools are so accessible within the libraries, they are often used for B-2-B projects without LIT's awareness. A B-2-B student stopped by the Marriott's weekly makerspace event and asked about 3D printing. The assistant head of Creative Spaces, a LIT member, was sitting in front of a printer at the time and walked him through the process as the model printed. It was a short print, and the student was able to make design changes on the fly and print several versions. Only when the LIT member saw the student at the B-2-B competition with the prototypes that they coproduced did he realize the prints were for a B-2-B project. While maximizing access to services is always a priority in libraries, not tracking what these prototyping tools are used for makes it difficult to determine when librarians are involved in the research and innovation process.

Conduct Outreach to Innovators

LIT relies heavily on interactions, both formal and informal, between LIT members and U of U partners. Developing relationships with these partners and being involved in the ongoing efforts of their departments is important to the growth of the LIT. Team members use LIT meetings to discuss new relationships and outreach efforts and brainstorm additional opportunities for partnerships.

With an increasing presence at B-2-B events and workshops, LIT decided to create two promotional materials to serve as reminders of LIT. The first promotional material was a flash drive. Since each B-2-B team is required to create a poster for the competition event, the LIT team purchased flash drives and loaded poster templates on them to support the creation of quality posters.

The second item was a trifold business card highlighting specialized innovation services, the librarians with that expertise, and contact information. These cards fit in a wallet or pocket, so they could be easily consulted when needed. With the expansion of innovation services, the trifold quickly became obsolete. As a more sustainable alternative, LIT's leadership designed a sticker that could be added to individual LIT members' personal business cards. These stickers include LIT's website and e-mail address and can be updated and reprinted quickly at a lower cost. This has proven to be a more effective method of promotion, as B-2-B teams can request a consultation with the individual they have been in contact with or be referred to LIT as a whole.

Each of the three libraries contributed a small amount of money to purchase additional promotional materials to be distributed at B-2-B kickoff and workshop events. A small subgroup of LIT worked with an advertising merchandise company to select and design these promotional materials. Given the available budget and the advice of the merchandise company, the subgroup selected lip balms and pens with the LIT logo and website.

Impact

At the close of the B-2-B competition each year, a report detailing information about the teams, the competition night, and the winners is written and distributed. The LIT leadership and members are consistently acknowledged for their support of the B-2-B competition. An article detailing how student teams worked with librarians throughout the competition appeared in one competition report, highlighting the role LIT plays in supporting B-2-B teams and encouraging future teams to consult with librarians for their success.[10]

Innovation is a significant and growing part of the culture at the U of U. Recognizing that different areas of the U of U have created their own innovation groups and wanting to combine their collective strength, an Innovation Ecosystem website highlights all of these innovation partners. The Innovation Ecosystem website brings together innovative groups, encouraging collaboration across the U of U. It also features resources that provide support to these groups, including LIT. LIT is proud to be recognized as an invaluable resource for innovation groups at the U of U.

LIBRARIAN PERSPECTIVES

This next section provides librarian perspectives on assisting innovators. The three case reports—patents, market/business, and biomedical—include examples of librarians partnering with B-2-B teams.

Case Report 1: Patents

Two of LIT's earliest successes occurred before it was officially formed and helped create the organizational momentum for LIT. These cases involve two U of U students. This example effectively demonstrates how cooperation between librarians and students creates networks that drive success.

One student participated in the B-2-B competitions in a variety of roles, including serving as the engineering chair and then health sciences chair, as a junior vice president, and finally as president in 2013. He worked with the patent and trademark librarian while on two different B-2-B teams to help determine the prior art for two areas of technology: one a cervical biopsy device and the other a pediatric diagnostic teddy bear. These two teams became award winners—the cervical biopsy tool won Grand Prize in 2011, and the *TeddyCare* won the Entertainment Arts and Engineering Award for 2013. While organizing the competition in August 2013 for the next academic year, this student leader Patrick Loftus wrote in an e-mail to a LIT member,

> One of the key components of any new innovation is a thorough understanding of prior art and the untouched intellectual property available to the inventor ... as a participant in [B-2-B], [the patent and trademark librarian's] knowledge and help were the fastest way to jumpstart our medical device. Without him, it would have taken us months of floundering to reach the same point that only took one night and a couple [of] meetings. The [B-2-B] program has directed my career as a future physician into a realm of innovation. Without [the patent and trademark librarian], I imagine my initial months of medical device development would not have been as exciting or optimistic, and as such, would not have had such a positive impact on me and my future.

The second case involves an undergraduate student in the Multi-Disciplinary Design and Entrepreneurship Program. While on a ski trip, he had a problem with his gear that sparked an idea for an improved ski pole. His design allowed the pole to be adjusted and customized easily and included a basic toolkit in the handle to allow skiers to make repairs on ski slopes. The idea turned his interests toward U of U classes on innovation, and he began to regularly seek research assistance from Marriott. After meeting with the patent and trademark librarian, he applied for a U.S. provisional patent in 2012 and completed a nonprovisional U.S. patent application within the U.S. Patent and Trademark Office's (USPTO's) one-year deadline. His complete U.S. patent was granted 28 months later.[11] These coordination efforts serve as an effective demonstration of how librarians help innovators turn their ideas into reality. The student later participated in a promotional video highlighting student innovation.[12] In the video, he describes how library resources are essential to product design, distribution, and patent protection efforts.

Case Report 2: Market and Business

Finding market and business information is difficult for B-2-B students. Library faculty and staff have expertise in the use of terminal resources, such as *Bloomberg*, *LexisNexis*, and *ThomsonOne*, as well as business databases, such as *Factivia* and *Mergent*. They also provide a deep understanding of business classification systems, such as the Standard Industrial Classification and the North American Industrial Classification System. Since B-2-B's inaugural year, additional marketing and business databases were purchased to support B-2-B.

The 2016 B-2-B grand prize winner contacted LIT for assistance on market research. Spencer Madsen stated in an e-mail to a LIT member on April 11, 2016,

> I contacted the Library to help find more specific market numbers regarding my baby monitor. Unfortunately it was really hard on my own to find anything that didn't require an expensive purchase of a market analysis. Thankfully, [two librarians] spent a couple hours with me searching. We didn't answer all my questions, due to the complexities of my specific market, but we did find some new information that was essential to depicting the size and direction of the market. That information was used to create an enticing market story, which helped the judges see the potential impact of my technology.

The main obstacle for this B-2-B student is that most of the current baby monitor manufacturers create baby monitors as a side project rather than as their main revenue stream. This makes it difficult to pinpoint the exact market size. By working together, the LIT members used several business databases to find a few tangential reports. The student used these for his B-2-B poster and discussion topics to demonstrate the value of the device to the B-2-B judges.[13]

LIT also worked with an innovator to locate competitive intelligence on two specific companies. These two companies offered software platforms that help to silence patient rooms and minimize alarms for health care professionals. Due to legal requirements and established work flows, these two issues are constantly recorded as problems; however, uncovering solutions is a challenge. A U of U faculty member wished to investigate two companies to determine which company would be the best partner or whether the U of U should seek its own software solution. This project required a variety of expert research. One LIT member discovered the companies' patents, and another investigated the biomedical literature for publications by the companies and their subsidiaries. Both searched business databases for the companies' financials and other reports. With these information resources, the U of U faculty member made an educated and informed decision for the next step of this project.

Case Report 3: Biomedical

B-2-B teams search the biomedical literature to verify that the technologies they are developing address an unmet clinical need or provide a novel solution. In order to effectively navigate this literature, B-2-B student teams regularly meet with EHSL librarians. As a former B-2-B student Chris Bowen stated to a LIT member in 2015, the health sciences librarians "helped leverage my access to the most recent literature published on the problem we were striving to solve . . . [and they provided] guidance [on] how to maximize our literature searches to find the breadth of information to accurately understand the clinical need and depth of studies already performed."

For most biomedical literature requests, EHSL faculty meet with one or two members from a B-2-B team and teach them how to search databases, such as *PubMed* and *Embase*. EHSL purchased access to *Embase* to expand EHSL's systematic review services (http://library.med.utah.edu/or/services/litsearch.php). Since *Embase* includes more medical device literature, it is a valuable tool for medical device researchers. Other databases and tools, such as *SPORTDiscus*, *CINAHL*, and FDA's *MAUDE Database*, may be relevant to the B-2-B teams, depending on their devices.

Some questions from B-2-B teams are unique. For example, one B-2-B team requested the average measurements of an adult human head. In this case, the answer was discovered in EHSL's print books rather than through online resources.

Consultations and interactions with B-2-B teams involving biomedical literature have covered a wide spectrum of topics. Medical code research is a popular topic and complex enough to require a separate page in LIT's *Innovation Guide*.[14] Medical codes include the *Current Procedural Terminology* code set, maintained by the American Medical Association and used to describe medical services, and the *International Classification of Diseases* codes, created by the World Health Organization for the diagnosis of medical conditions. B-2-B teams use these code sets to investigate topics, such as the market size and national cost of particular medical conditions and services.

For example, if a team wants to develop a device that works with computerized axial tomography (CAT) scans, they could use the relevant codes to discover the number of CAT scans performed in the United States for a specific year. This information would provide a general U.S. market size. Furthermore, if this team wants the device to be used in a developing country, they can use the code and data set to gather additional information, such as a breakdown of CAT scans performed in the United States based on the size of a hospital or clinic. The data on smaller health care settings could provide a proof of concept before deploying medical devices in less developed countries.

While planning market entry strategies for their devices, B-2-B teams ideally want to find a preexisting code that covers their devices rather than going through additional steps to create a new code. If a B-2-B team's device requires a new code, this can be a substantial hurdle to cross before the new technology can be integrated into a health care setting.

Biomedical literature and health sciences librarian assistance provide the foundation for exploring many B-2-B team ideas. A former winning B-2-B student, Tim Pickett, stated to a LIT member in 2015,

> The library and its resources have been an invaluable resource . . . [they are] really positive and encouraging towards student innovation and on multiple occasions taught us how and where to find the information we needed . . . [they] were key in understanding the disease state fundamentals which was our springboard to creating a portable and cost effective therapy for the developing world that prevents precancerous lesions from ever turning into full blown cervical cancer.

COMPETITION STUDENT LEADER PERSPECTIVES

Each year, new student leaders are recruited to direct the B-2-B competition, organize the year's events, and provide guidance to new student participants. In this section, two student leaders provide their perspectives regarding how libraries can best team up with innovators.

B-2-B Competition Summary and History

Since its inception and inaugural year in 2010, B-2-B has mentored 624 participants on 143 teams that have invented 147 medical devices, filed 108 patents, and launched 38 limited liability companies (table 10.2). This is possible through the leadership of students in medicine, business, and engineering, as well as physicians, mentors, administrative staff, and, of course, librarians. The competition has also benefited from sponsors in the local business and banking community. This collaboration between academic and industry sponsors allows the competition to reach a greater number of students and have a much broader impact.

Table 10.2. Bench to Bedside Statistics.

	2016	2015	2014	2013	2012	2011
Participants	112	116	189	74	57	76
Teams	32	24	42	18	14	13
Devices developed	32	24	43	20	14	14
Provisional patents filed	18	14	16	14	13	12
Utility patents filed	4	2	5	5	4	1
Limited liability companies formed	11	3	12	7	4	1

Building the Future Innovative Academic Library

In an ideal world, having access to all of the tools necessary to take an idea all the way to a thriving company located in a single location would be incredible. Unfortunately, it is simply not feasible. Academic centers and businesses across the country face this problem no matter how much funding they have. Academic centers advance research. Hospitals provide the best patient care and medical treatments. Industry-leading manufacturers understand new technology, building, and design. But having access to all of this expertise in a single location and making it accessible to students not only is unlikely but also could be a logistical nightmare. Opportunities for networking, though, are key to innovation. Having a dedicated collaboration space, especially at a centrally located library, is ideal for such collaborations. One of the 2016 B-2-B leaders said, "The library cannot do everything, but it can connect to everything. Being able to integrate with what others are doing keeps innovation impactful, inexpensive, and timely." To this end, in figure 10.1, the B-2-B student leaders identified the top seven areas that should serve as pillars in building a collaborative academic library of the future that serves innovators.

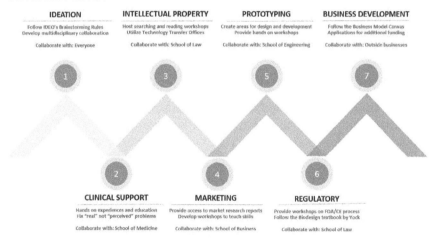

Figure 10.1. Seven Areas of Innovation.

Ideation

Ideation is the process of creating or developing a new idea. It is an essential component of innovation, but there is no set structure or method for engaging in the ideation process as individuals generate ideas in a variety of ways. Although ideation cannot be forced, certain environments and techniques have proven to be helpful in facilitating creative thought processes.

Students find it helpful to have a private place where they can openly discuss ideas. Many are visual learners and can better visualize ideas using whiteboards, SMART Boards, and tinkering with physical objects (e.g., pipe cleaners, clay, and building blocks). Libraries can have designated ideation areas with carts full of play items to encourage student innovation as the EHSL has done with its Ideation Studio. As these spaces create an ideal setting for idea generation, it would also be beneficial to provide tools, such as IDEO's seven tips to support effective brainstorming (https://challenges.openideo.com/blog/seven-tips-on-better-brainstorming), to help guide innovators on a productive idea generation path. When possible, it is also useful to include replicas of equipment used in clinical settings. The CMI, for example, has asked local hospitals for expired medical devices and equipment to help students develop improvements or new medical devices. Interacting with these devices increases student understanding of what is needed in health care settings and allows them to imagine creative solutions for products already being used.

Multidisciplinary teams are also helpful in ideation. B-2-B encourages students from different colleges, including engineering, business, medicine, law, and design, to come together and use their different backgrounds and skill sets to identify a clinical problem, create a solution for that problem via a medical device/technology, and develop a strategy to commercialize it. With students coming from such disparate areas, it is important to have a collaborative, neutral space in which to brainstorm. Libraries naturally lend themselves to these requirements and can significantly enhance innovators' experiences by investing in these kinds of spaces (chapter 3).

Clinical Support

Ideas, feedback, and support from health care professionals are vital in medical innovation. Practicing clinicians are experts in their respective fields and know the problems through hands-on experience. Students should work directly with health care professionals to identify and address clinical needs. The U of U's Department of Bioengineering offers both undergraduate and graduate courses that give bioengineering students the opportunity to do observations in the operating room and other clinical settings. These observations give students firsthand exposure to clinical issues and the opportunity to collaborate with health care professionals who provide input as to whether the medical innovation idea is useful and viable.

Even with the expert advice of clinical faculty and staff, student innovators need to identify evidence for why they should pursue particular medical innovations. Clinical research reports are particularly valuable for this, though they can be difficult to locate. Libraries provide access to databases, such as *PubMed* and *Embase*, which may contain some clinical research reports; however, these reports may also be in the gray literature. Student innovators rely on librarians' proficiency in searching the literature in more obscure places to help them find existing research relevant to their innovations.

Intellectual Property

Intellectual property is another crucial component of medical innovation that must be considered by student inventors and entrepreneurs. The U of U has a USPTO-certified patent search librarian dedicated to helping students with patents. This librarian schedules appointments with students to teach them about patents and how to use the different patent search engines. Searching the prior art (existing patents) should be one of the first steps taken by student innovators. Having a patent librarian is an extremely valuable resource that makes this step easier.

In addition to searching prior art, innovators are also required to file their own patents. The CMI hires law students who have experience with patents to work with teams developing medical technologies. These law students work with the teams to understand the technology they want to create and then help them draft and file a provisional patent. Provisional patents protect the inventors' intellectual property for one year from the date of filing, giving inventors the opportunity to validate their ideas and find funding to pursue the technologies. Academic libraries should consider collaborating with law students to help student inventors establish and protect their intellectual property.

Marketing

Marketing is traditionally thought of as the most important factor in acquiring a customer. A tool called the *four Ps of marketing*,[15] which emphasizes price, place, product, and promotional strategy, is often used to develop a marketing plan. It is important for innovators to have access to market research reports to understand how a new product or service fits within the existing industry. Unfortunately, these documents are quite expensive, but it is vital for a company to understand the four Ps. Libraries are an ideal place to bridge this information gap for student innovators. Some groups that produce these reports make academic licenses available to help reduce the price burden for access to this information. Librarians can collaborate with campus business schools to purchase licenses to these reports and gain training in reviewing market research reports and developing strategy related to specific markets.

Prototyping

While not every library has the space or the funding for laboratories and prototyping areas, as discussed in chapter 6, inexpensive prototyping equipment allows invaluable opportunities for hands-on innovation. One of the most important and inexpensive tools for libraries to have is a 3D printer. A fused deposition modeling printer, for example, is essentially a high-quality glue gun that takes a spool of plastic filament and threads it into a heated extruder nozzle. This relatively inexpensive piece of equipment allows students and faculty to have rapid prototyping of their innovative ideas.

This is another area in which libraries can have much success in collaborating with other departments. Colleges of engineering, fine arts, and even architecture may have a variety of equipment useful for prototyping and building new innovations. Building on the expertise of these entities serves libraries well in building new partnerships and helps students and other innovators develop their ideas.

Regulatory

Often, new innovations must go through a regulatory body before entering the market. In the medical field, this is typically the U.S. Food and Drug Administration (FDA). As understanding the intricacies and importance of the FDA is no small matter, librarians can provide workshops on how to search and understand the FDA website (http://www.fda.gov/Medical Devices). These searches identify predicate (similar) devices already approved by the FDA, assisting innovators in determining testing procedures and applicable standards.

In the book *Biodesign: The Process of Innovating Medical Technologies*,[16] Yock and colleagues provide an excellent overview of the medical innovation process. Specifically, they discuss the various pathways through the FDA and strategies for navigating each. This book (and others like it) is particularly useful for libraries interested in working with innovators who need guidance on the process of producing medical devices.

Business Development

The final and most important step in innovation development is taking an idea from the conceptual stage into an actual business. It is, by far, the most difficult step because it requires the greatest amount of funding, and, unfortunately, this is the stage in which most ideas fail. Through their campus partnerships, libraries can collect and distribute information about competitions, grants, and connections to incubators, angel investors, and even venture capitalists interested in supporting student innovation. Many of these resources are cultivated through websites such as *Gust* (https://gust.com),

Entrepreneur (https://www.entrepreneur.com), and *Startup Compete* (http://startupcompete.co). Librarians may also partner with schools of business to learn more about business development and funding opportunities.

While no single place can contain all of the information, resources, and expertise needed for students to work through the process of ideating, prototyping, and developing a business plan for their innovation, librarians can be important partners in directing them to what they need. By creating neutral spaces for student teams to meet, cultivating expertise in searching for specific kinds of development information, and identifying and connecting collaborators across universities and throughout the community, libraries and librarians become invaluable assets for innovators.

CONCLUSION AND FUTURE DIRECTIONS

Academic libraries are in a unique position to partner with university innovators and entrepreneurs. Not only can library faculty and staff connect innovators to different information resources and other experts, but libraries can offer neutral meeting spaces and unique services. Libraries are an excellent resource for innovation and entrepreneurship at universities, as they can offer needed resources, equipment, training, and software for students' academic success and future careers. At the U of U, the LIT always receives accolades from B-2-B teams for helping them improve the quality of their projects. This is best demonstrated by the fact that 11 out of 12 of the winning B-2-B teams in 2015 worked with and recognized LIT members. One 2016 team member, Scott Anjewierden, even remarked to a LIT member, "I wish I would have e-mailed the library sooner, they were so helpful!"

As LIT looks to expand its role in the future, team members continually work to identify new partnerships and areas for collaboration. In the fall of 2016, the Lassonde Entrepreneur Institute opened a residential community called Lassonde Studios. Lassonde Studios is designed for student entrepreneurs who *live, create, and launch* products from shared *garage* spaces. LIT identified several outreach opportunities, including offering workshops and holding virtual or in-person office hours in the Lassonde Studios.

There are numerous other innovation activities and groups at the U of U, and LIT has only scratched the surface of its capabilities. LIT serves as an invaluable partner to university innovators not only through the spaces and services offered by the libraries but also in connecting innovators with one another. For example, EHSL and Marriott partnered with Technology and Venture Commercialization to cosponsor a lecture on making disruptive innovations mainstream.[17] While institutions can often be siloed, the multilibrary makeup of LIT creates an understanding of innovation across the U of

U, allowing collaborators from law to business to the health sciences to be connected and informed through LIT's efforts.

NOTES

1. J. Willard Marriott Library, "Our Mission, Vision and Values," 2016, http://lib.utah.edu/info/mission (accessed June 24, 2016).
2. Spencer S. Eccles Health Sciences Library, University of Utah, "About Us," 2016, http://library.med.utah.edu/about (accessed June 24, 2016).
3. Libraries Innovation Team, University of Utah, "Innovation: Welcome," 2016, http://campusguides.lib.utah.edu/innovate (accessed June 24, 2016).
4. Tallie Casucci, "Judges Needed for Games4Health Challenge," *Spencer S. Eccles Health Sciences Library Blog*, 2015, http://library.med.utah.edu/blog/eccles/2015/11/24/judges-needed-for-games4health-challenge (accessed October 21, 2016).
5. Tallie Casucci, "LITE Awards $2000 to B2B Innovation Team," *eSynapse: Newsletter of the Spencer S. Eccles Health Sciences Library*, 2015, http://library.med.utah.edu/esynapse/2015-april-june-vol-30/lite-awards-1000-to-b2b-innovation-team (accessed June 24, 2016); Tallie Casucci, "Library Innovation Team Award Winner at Bench-2-Bedside," *eSynapse: Newsletter of the Spencer S. Eccles Health Sciences Library*, 2016, http://library.med.utah.edu/esynapse/2016-april-june-vol-31/library-innovation-team-award-winner-at-bench-2-bedside (accessed June 24, 2016).
6. Alexander Osterwalder and Yves Pigneur, *Business Model Generation: A Handbook for Visionaries, Game Changers, and Challengers* (Hoboken, NJ: Wiley, 2010).
7. Libraries Innovation Team, University of Utah, "Innovation: Business Model Canvas," 2016, http://campusguides.lib.utah.edu/innovate/bmc (accessed June 24, 2016).
8. Tallie Casucci, Timothy Pickett, and Jean P. Shipman, "The FabLab Is Fabulous!," *eSynapse: Newsletter of the Spencer S. Eccles Health Sciences Library*, 2015, http://library.med.utah.edu/esynapse/2015-april-june-vol-30/the-fablab-is-fabulous (accessed June 24, 2016).
9. Tallie Casucci, "New FabLab Tools: Objet30 Prime and ISO Standards," 2015, *eSynapse: Newsletter of the Spencer S. Eccles Health Sciences Library*, http://library.med.utah.edu/esynapse/2015-july-september-vol-30/new-fablab-tools-objet30-prime-and-iso-standards (accessed June 24, 2016).
10. Center for Medical Innovation, University of Utah, "Campus Librarians: Information + Innovation = Success," *Competition Report 2015 Bench to Bedside*, 2015, 8, http://library.med.utah.edu/e-channel/wp-content/uploads/2016/04/ReportB2B2015.pdf (accessed June 24, 2016).
11. Alexander William Carr, "Multifunction Ski Pole," *Google Patents*, 2015, http://www.google.com/patents/US9101818 (accessed June 24, 2016).
12. Alexander William Carr, "'All U Need' for Business Student, Entrepreneur, and CEO of Char Poles," video, produced by J. Willard Marriott Library, University of Utah, November 26, 2015, https://youtu.be/hFJgoLLImEI.
13. Lassonde Staff, "Bench-to-Bedside Grand-Prize Winner Could Prevent Infant Death," *Lassonde Entrepreneur Institute Press Releases*, 2016, http://lassonde.utah.edu/bench-to-bedside-grand-prize-winner-could-prevent-infant-death (accessed October 21, 2016).
14. University of Utah Libraries, "Innovation: Medical Codes," 2016, http://campusguides.lib.utah.edu/innovate/medicalcodes (accessed June 24, 2016).
15. William Perreault Jr,, Joseph Cannon, and E. Jerome McCarthy, *Basic Marketing: A Marketing Strategy Planning Approach* (New York: McGraw-Hill/Irwin, 2013).
16. Paul G. Yock, Stefanos Zenios, Josh Makower, et al., *Biodesign: The Process of Innovating Medical Technologies*, 2nd ed. (Cambridge: Cambridge University Press, 2015).
17. Michael Thomas Eckhardt, "Crossing the Chasm: How to Successfully Bring New Disruptive Innovations to Mainstream Users and Customers," 2015, e-channel, http://library.med.utah.edu/e-channel/portfolio/crossing-the-chasm (accessed June 24, 2016).

Chapter Eleven

Innovation Space Drives Need for Librarian Expertise

Jean P. Shipman and Tallie Casucci

BACKGROUND

In 2012, the Spencer S. Eccles Health Sciences Library (EHSL) offered office and meeting spaces to the Center for Medical Innovation (CMI) at the University of Utah (U of U). These spaces expanded as more innovation-related entities became residents of the EHSL Garden Level (chapters 6, 8, 9, and 12). This renovated medical innovation and collaboration space became known as the Synapse (chapter 3).

As EHSL faculty and innovators interacted within the Synapse, it became apparent that librarian expertise would accelerate the application of knowledge. The EHSL director discussed with the CMI executive director how a librarian would benefit the CMI programs. A professional librarian was hired on a temporary basis to test the concept and to complement and enhance the innovation processes taking place within the Synapse.

INNOVATION LIBRARIAN

In early 2014, a research assistant was hired for a year to develop effective working relations with the Synapse occupants and to test the necessity for a dedicated innovation informationist. Soon, this *innovation librarian* (IL) became an accepted professional partner of the students, as she inserted her expertise into many innovation initiatives (chapter 9). Within six months, the EHSL director decided to hire a full-time tenure-eligible faculty position. A national search was conducted for the new IL, resulting in the hiring of the temporary research associate.

The IL's responsibilities include (1) supporting students, faculty, and staff working on medical gaming, innovation, and medical entrepreneurism; (2) mentoring innovation fellows and other part-time employees; (3) participating on teams responsible for literature searching, general reference, education, and outreach; (4) planning and coordinating innovation and gaming-related events; and (5) promoting EHSL's value via social media. The IL position became part of the EHSL's Public Services Division for several reasons, including to build the cadre of librarians trained to perform consulting and extensive literature reviews and to collaborate with other librarians to develop and maintain professional skills. This position's placement has proven to be valuable for many reasons, as work can be distributed at the appropriate employee level. For example, clerical tasks are assigned as *special projects* to part-time Public Services staff under the IL's direction.

The IL's educational background and continuing professional development are essential to the job. Fortunately, the IL's education background includes a bachelor of arts in a health sciences field, a business certificate, and a master's degree in library and information science. The IL is a combination of a health sciences librarian, business/entrepreneur librarian, and prior art/patent librarian. To remain current within these librarian profession areas, the IL is a member of different professional organizations, monitors several listservs, and reads various journals and blogs. The IL is a health sciences librarian first and foremost, but knowledge of business, marketing, engineering, gaming, and patent resources is essential. With this unique blend of skills and knowledge, it is difficult to find focused conferences and training opportunities. Currently, the IL alternates between attending various conferences, such as the Medical Library Association, Special Libraries Association (http://www.sla.org), VentureWell, and Digital Games Research Association (DiGRA).

DIRECTOR FOR INFORMATION TRANSFER

In January 2014, the CMI executive director appointed the EHSL director as CMI's director for information transfer (DIT). As both the EHSL director and CMI DIT, this individual sees the needs of each entity and attempts to bridge these needs. The position has the advantage of a panoramic view of daily responsibilities and tasks of both entities and believes that duplication of effort should be eliminated via Lean principles.[1] With public space use and administration knowledge, the EHSL director guides the development of joint policies and procedures and can directly tackle facility issues, such as building access, room scheduling, facilities management, and security. She is aware of U of U policies and procedures and serves to assist with purchases and interactions with telecom and data networking professionals and offers financial and accounting advice.

The DIT and associate dean of the U of U J. Willard Marriott Library held focus group sessions with past U of U innovation students to understand their information needs. They asked questions concerning when information was needed and how the interviewees obtained such information. As a result, the DIT and associate dean appointed a Libraries Innovation Team (LIT) (chapter 10) to work with the CMI programs and other U of U innovation activities.

The DIT initiated an innovation-related platform, named e-channel (chapter 13), designed to collect the outputs of not only innovators from the CMI but also innovators in education, research, and health care at the U of U and beyond. Its goal is to transfer the knowledge that is generated by innovators to others. The DIT envisions e-channel becoming the repository for anyone wishing to locate relevant innovation and entrepreneurship information. It would be the ideal location for U of U Innovation Ecosystem partners to host developed content and output so that it is preserved and discoverable.

Another role of the DIT is to encourage individuals, U of U entities, and professional organizations and associations to contribute their conference proceedings and content to e-channel to help preserve this vital information. A strategy for capturing gaming products that are created each semester by The GApp Lab (chapters 8 and 12) is an example of content that needs to be included in e-channel. Articulating the vision for the further development of e-channel is a key responsibility of the DIT.

As a result of working with e-channel and innovation-related professional associations, it became clear to the DIT that an ontology of innovation was needed. The ontology would provide consistent metadata for similar themes and topics to describe posters, videos, podcasts, presentations, and so on. Standard terminology to describe innovation-related content is being compiled. Where appropriate, a relationship hierarchy is being created for the terms. After final refinements and testing, the ontology will be released to the public.

U OF U INFORMATION TRANSFER EXAMPLES

Both the DIT and the IL partner directly with U of U entities and other innovators from national and international professional organizations. Information transfer projects implemented with U of U entities are described below.

CMI

The DIT and IL work with CMI staff to enhance their programs and improve their work flows. There are two work flow improvements the IL helped to implement. First, the IL taught Microsoft Excel and Microsoft's mail merging features to CMI's associate director for education (ADE). Prior to learning about mail merge, the ADE was e-mailing each person individually and simply changing the salutation for each e-mail. Her contacts were in disorganized electronic documents, saved e-mails, and handwritten lists. After teaching the ADE about these products, she now easily organizes CMI's contacts and e-mails personalized announcements. This drastically reduces the time required to create and send announcements to various program participants (e.g., student competitors, mentors, and faculty innovators). In fact, the ADE recruits new competition judges using this efficient system. Each new judge inquiry includes a sentence about the person's expertise and his or her relevance to the competition.

The second improvement was restructuring and simplifying CMI reimbursements and its payment request system. Prior to the IL's involvement, the ADE spent days tracking down receipts, calling program participants, and completing various forms to reimburse competition participants for purchases. The IL suggested writing a *CMI Reimbursement Handbook*. This handbook includes an overview of the available reimbursement types and instructions for completing CMI's associated forms. This distributed the work to CMI participants. With the *CMI Reimbursement Handbook*, participants can learn about the reimbursement process and pick the reimbursement or prepaid method that best matches their situations. The handbook's forms are in PDF format, allowing participants to type the required information, minimizing handwriting discernment challenges. After the success of the handbook, the ADE created other forms to improve work flows, such as travel reimbursement forms. With these forms, all the required and needed information is readily available for the ADE.

Additionally, to facilitate reimbursements, a few of the EHSL part-time Public Services staff were trained to process these requests. This means taking information from the *CMI Reimbursement Handbook* forms and entering the data into the U of U's Accounts Payable required forms. The completed forms are printed and returned to the ADE to sign and submit to Accounts Payable. Program participants can collect their reimbursement checks at the EHSL Front Desk at their convenience. By utilizing the Front Desk staff, the ADE has more time to plan CMI's program events rather than being tasked and overwhelmed with simple data entry and paperwork. Also, the ADE has more flexibility with her schedule, as students are not interrupting her to collect their competition reimbursement checks.

Bench to Bedside Competition

The main program organized by CMI is the student competition Bench to Bedside (B-2-B). B-2-B is an annual student team medical device development competition. Multidisciplinary student teams receive $500 of prototyping funds to develop a device to address an unmet clinical need. The IL works with these student teams to direct and teach them about finding relevant biomedical literature, statistics, and market data for their devices. The case reports in chapter 10 highlight the IL and other LIT members' roles with assisting B-2-B teams.

At each annual kickoff event for the B-2-B competition, the DIT introduces students to the LIT. During this brief introduction, the DIT highlights the expertise of the LIT members and the success of former B-2-B teams who consulted them. In 2015, 11 out of the 12 winning B-2-B teams worked with a LIT member.[2] In 2016, 65 percent of all teams and 70 percent of the winning teams utilized such expertise.[3]

Innovation Corps

Innovation Corps (I-Corps), offered by the National Science Foundation, "is a public-private partnership program that teaches grantees to identify valuable product opportunities that can emerge from academic research, and offers entrepreneurship training to student participants."[4] The CMI applied for and was awarded a three-year grant to offer this program annually from 2014 to 2017 as an I-Corps site. This faculty-targeted program consists of a didactic lecture-based curriculum that features speakers from the U of U faculty, industry partners, and entrepreneurs from the Salt Lake City community. Lecture topics cover the Business Model Canvas (BMC) components (https://en.wikipedia.org/wiki/Business_Model_Canvas 2016), such as the value proposition of the proposed product or service, revenue model, regulatory considerations, and capital acquisition strategies. The BMC framework is based on the book *Business Model Generation*.[5]

Most I-Corps sites use a Web-based BMC program called *LaunchPad Central* (https://www.launchpadcentral.com). *LaunchPad Central* allows teams to develop their business hypotheses, product ideas, and designs within the BMC framework. Each BMC component in *LaunchPad Central* includes questions and videos to guide teams to the appropriate information. As teams work through the 15-week U of U I-Corps program, they outline strategies and verify assumptions in *LaunchPad Central*. As new content (e.g., assumptions, customer interviews, etc.) is added, the pivots, or changes in strategy, are tracked and documented. *LaunchPad Central* is a great tool for tracking interviews and communicating with fellow team members. Teams transcribe their notes from customer interviews so that they can refer back to interview findings. Team members can add comments to all entries and exchange ideas and insights within the shared work space. Additionally, I-Corps instructors and mentors can comment on teams' progress.

The CMI executive director asked the DIT and IL to operate and teach the *LaunchPad Central* software to the 2015–2016 and 2016–2017 faculty cohorts. I-Corps teams, instructors, and mentors are registered within the software. At an I-Corps meeting, the DIT and IL teach *LaunchPad Central*, highlighting its value and many features with an example product and team. The IL and DIT monitor the teams' progress during the cohort. Through this training, the DIT and IL gain a unique perspective to the I-Corps teams' information needs. They also attend the lectures to answer any questions regarding *LaunchPad Central* usage or other information queries. EHSL videotaped the lectures for the first cohort for later reference by all teams.

The I-Corps teams have posed a few information queries to the IL. One team needed help finding relevant biomedical literature and competitive intelligence on similar technologies and companies as their concept. The IL supplied several *PubMed* (http://www.ncbi.nlm.nih.gov/pubmed) and *Embase* (http://www.embase.com) search strategies and links to information on similar technologies and companies. Another I-Corps team member consulted the IL as she completed documentation for a clinical evaluation report of her medical device. She needed to develop search strategies that included supportive evidence for the validity of her team's product.

Business Model Canvas Unique Applications

After working with the I-Corps teams and the *LaunchPad Central* software, the DIT and IL developed two unique applications for the BMC framework. The IL paired relevant library databases and university resources to the different steps of the BMC for B-2-B teams. Since U of U affiliates have access to numerous resources, this mash-up tool alleviates some information and resource overload. It directs B-2-B teams to relevant information resources so that the teams can target their evidence searching to the most appropriate tools. This mash-up tool is linked on the *Innovation LibGuide*,[6] developed by LIT, and the PDF version is shared with B-2-B teams. The tool is a starting place for B-2-B teams as they prepare for the annual competition capstone event. The tool was highlighted at the Open 2016 VentureWell conference so that others can apply this strategy.[7]

The second application utilizes the BMC framework to develop an individual's professional career as a *product*. There are many similarities between designing a product and designing or strategically planning one's career. A BMC was completed with career-related items, and appropriate library classes and information resources were aligned to the various BMC steps. This template has turned into a one-credit course of seven lectures that will be taught in the summer of 2017 to U of U College of Pharmacy students as a pilot. An abridged version is being taught as an EHSL workshop to anyone. If successful, the course will be offered to all health sciences students and possibly as a part of the U of U's InterProfessional Education Program, as this information is valuable to all health sciences students.

BioInnovate

> Another program, BioInnovate, offered by the U of U Innovation Ecosystem, aims to provide a comprehensive biomedical device design training program through the use of a multidisciplinary, hands-on teaching approach in classroom, clinical, and laboratory settings. Students from engineering, medicine, business and law backgrounds . . . [are] trained in clinical problem identification, medical device innovation, and commercial translation; all within the regulatory framework of the FDA. Students . . . immerse themselves within clinical environments and learn to observe procedures, the utilization of medical devices, and interact with patients and clinicians to uncover unmet clinical needs. Students . . . translate these unmet needs into medical device concepts that will be refined for commercial potential. Once final concepts have been generated, student teams . . . further develop these ideas into testable prototypes and develop business plans while operating under the regulatory framework of the FDA.[8]

This yearlong graduate course is taught at different locations, including the Synapse (chapter 3). BioInnovate students can ask the IL for information assistance.

In 2015, the DIT offered to evaluate BioInnovate since it completed its fifth year. The IL created a step-by-step process for an EHSL part-time Public Services staff member to perform a preliminary investigation of former BioInnovate student achievements. After the staff member completed the preliminary investigation, the IL discovered additional information and created an initial report. The BioInnovate scoping review report includes graphs of initial data and future plans. The DIT and IL met with the BioInnovate director to discuss next steps and determine other metrics and information to include in the report to highlight the program's significance. The IL contacted the former BioInnovate students to assess the program to create an official state-of-the-union report on BioInnovate's five-year history and impact. While this is a new partnership, future opportunities with BioInnovate include guest lecturing, integrating more program assessments, and tracking BioInnovate projects' outcomes and successes as well as its failures.

Technology and Venture Commercialization

The Technology and Venture Commercialization (TVC) serves as the U of U's technology transfer office. TVC staff contact the IL when working with health or medical start-ups. For example, TVC needed a list of specific health professionals to create a marketing campaign for a start-up. The IL referred them to several resources that could assist them, such as the *ABMS Directory*, professional organizations, and the Utah Division of Occupational and Professional Licensing.

Annually, the EHSL hosts an *InfoFair* to showcase different technologies and their information applications (http://library.med.utah.edu/or/infofair). In the fall of 2015, EHSL and TVC partnered to host a joint *Tech Tuesday* and *InfoFair 2015* combination event (http://library.med.utah.edu/or/infofair/infofair2015). *Tech Tuesday* is a recurring TVC event, held during the academic year, to host speakers who discuss a variety of entrepreneurial topics for U of U attendees as well as those from industry and the local community (http://www.tvc.utah.edu/events.php). The 2015 joint session invited Michael Eckert of the Chasm Institute to speak about *Crossing the Chasm: How to Successfully Bring New Disruptive Innovations to Mainstream Users and Customers* (http://library.med.utah.edu/e-channel/portfolio/crossing-the-chasm). It was one of the most popular *Tech Tuesdays* ever and the first time TVC cohosted this event with another entity.

Medical Digital Therapeutics at the U of U

There are two medical digital therapeutics (MTD) development programs at the U of U. The *Games4Health Challenge* is an international competition where university student teams design a wire-frame or beta version of a game related to health. The second group, The GApp Lab, is a graduate student game development lab. Chapter 9 covers several examples of the IL's work with these two groups. Chapter 13 presents how the EHSL works with these two MTD entities to capture their intellectual outputs to disseminate them via e-channel.

Virtual Reality Course

An example of the IL's work with MTDs was coteaching a special topics course on serious games. In the spring of 2016, the IL cotaught this class with an Entertainment Arts and Engineering (EAE) Program faculty member who is also the director of The GApp Lab. This graduate seminar class focuses on virtual reality (VR) within the realm of serious games. The course is divided by serious game topics, such as education, training, medicine, and military games. Each week's preclass work includes two scholarly readings, a popular press piece (video, news article, or podcast), and a discussion forum. Student leaders facilitate in-class discussions about the topics and associated readings; moreover, they bring a VR game for others to experience.

The final class project is creating a VR game. Unlike other EAE classes, the instructors stipulate the students' final projects be evidence based rather than just a random fun idea. The midterm presentation includes an annotated bibliography of materials that supports final VR game ideas. Many of the 2016 class students utilized the IL's expertise as they researched ideas for their final projects. At the end of the semester, the students presented their VR games at the EAE Program's public event, *EAE Day* (http://eae.utah.edu/events).

The initial VR serious games class was well received by the students. Next time, the instructors plan to integrate guidance on searching the research literature as a lecture. In preparation for upcoming semesters, the EAE faculty and IL have discussed how to teach future classes together, including a focused MDT class and a serious games writing class. Eventually, EAE plans to offer a serious games track for students who intend to pursue a career in creating serious games. If this happens, there will be many opportunities to integrate the IL's research expertise into the track's curriculum.

Colleague

Working with The GApp Lab director and a team of students, the DIT helped to create a researcher connection website called *Colleague*. This tool was commissioned by the U of U's Health System Innovation and Research Program (http://healthsciences.utah.edu/hsir/index.php). The *Colleague* tool, built to reflect internal U of U's faculty profile information, enables researchers to locate other researchers based on expertise and interests. The tool enables and facilitates new connections among different personnel at the U of U, not just the Health Sciences. It is a simple design, similar to the Tinder app. *Colleague* presents a person's biography and research interests, with methods for *favoriting* and contacting identified personnel.[9]

Drawing Health/Doodle Health

Health-related pictures, or pictographs, are frequently used to denote medical instructions to patients. Designing these pictographs can be a challenge since many are not culturally appropriate or understandable. The U of U School of Medicine's Department of Biomedical Informatics (http://medicine.utah.edu/bmi) funded a GApp Lab team to create a gaming app. This game, *Drawing Health* (http://drawinghealth.bmi.utah.edu/drawinghealth/home.php), allows vulnerable Utah populations to either guess the meaning of a pictograph or draw a pictograph for a health topic. If the gamer draws a new pictograph, that pictograph is added to the collection for others to guess its meaning. The game allows different populations to compete against each other. A public version, *Doodle Health*[10] (http://doodlehealth.com), allows anyone to contribute, not only vulnerable Utah populations. The DIT and the EHSL consumer health librarian participated on the design team to create the game. Additionally, the consumer health librarian took the game to local health fairs and events to garner community and consumer testers. Although the game is still in development, it already offers an open-source library of health-related pictographs for others to utilize (http://drawinghealth.bmi.utah.edu/doodlehealth/gallery.php).

Innovation Life Cycle

A final project that highlights the DIT and IL's roles with the entire U of U Innovation Ecosystem is creating an *innovation life cycle*. This ongoing project utilizes the DIT and IL's knowledge of innovation-related entities, resources, and services, both internal and external to the U of U. The innovation life cycle has four components: ideation and concept development, prototyping, product, and commercialization. Within each innovation component, relevant resources and services are provided for easy identification and access.

PROFESSIONAL ORGANIZATIONS INFORMATION TRANSFER EXAMPLES

There are two professional organizations' annual conferences that the DIT and IL attend and assist with transferring information among their organizations' members. The first is VentureWell, an organization that supports university entrepreneurial faculty, staff, and students in the United States. The second organization is DiGRA. "Founded in 2003, DiGRA is the premiere international association for academics and professionals who research digital games and associated phenomena."[11] Examples of the DIT and IL's roles and experiences with these professional organizations are described.

VentureWell

VentureWell, originally the National Collegiate Inventors and Innovators Alliance, was formed 20 years ago. VentureWell began offering grants to faculty to start programs in technology entrepreneurship, particularly ones that focused on the development of *E-Teams*. *E-Teams* are groups of students, faculty, and advisers working to commercialize a novel idea. VentureWell funds individual *E-Teams* from the entrepreneur courses and programs to help them bring their inventions to market. "VentureWell has grown rapidly to include a membership of nearly 200 colleges and universities from across the U.S., engaging thousands of undergraduate and graduate student entrepreneurs each year.... [It is] proud to be the leader in funding, training, coaching and early investment that brings student innovations to the world."[12] In addition to offering seed funding for *E-Teams*, VentureWell is an active partner with the National Science Foundation I-Corps program. It sponsors or supports other innovation-themed programs, including *Accelerating Startup Partnership and Investment Readiness*, the U.S. Department of State's *Global Innovation through Science and Technology* initiative, *University Innovations Fellows*, and the *Pathways to Innovation Program*.

VentureWell sponsors an annual conference for entrepreneurial students and faculty from universities across the country to share ideas through presentations, workshops, and posters and to learn from expert speakers (http://venturewell.org/open). U of U affiliates, including the DIT and IL, started attending this conference in 2014, and they have given presentations and contributed posters about the programs offered at the U of U.[13]

Starting in 2014, the DIT convinced VentureWell staff that their conference proceedings should be gathered and permanently stored to avoid losing access to this rich content. Collected conference proceedings are available via e-channel (chapter 13). EHSL staff assign metadata to the conference proceedings to facilitate easy discovery via e-channel's search engine. As content from past conferences disappears from the VentureWell website, having permanent access to such content offered by e-channel is invaluable (http://library.med.utah.edu/e-channel/portfolio/venturewell-open-conference). This is an example of how a library and a national association can collaborate to gather innovative content in one location for convenient access.

DiGRA

As stated earlier, DiGRA is an international association for game researchers. DiGRA has hosted annual meetings since 2013 and biannual meetings prior to that time. In 2014, the DiGRA conference was held in Utah. Due to its proximity, the DIT and IL attended the conference and presented with others from the CMI about their partnership and activities.[14] By meeting DiGRA members and listening to their presentations, the DIT and IL learned about how information plays a role with game development, especially serious game development. Many ideas for how the EHSL faculty could help DiGRA members surfaced. One key idea was to include conference proceedings in e-channel. Previously, conference proceedings were posted on DiGRA's website. An EHSL information professional assigns metadata to the conference proceedings. This allows e-channel to increase discoverability of the conference proceedings compared to DiGRA's website (http://library.med.utah.edu/e-channel/portfolio/digra-conference-material).

In 2016, the DIT attended the first joint conference with members from DiGRA and Foundations of Digital Games (FDG) (http://digra-fdg2016.org). FDG's "conference series seeks to promote the exchange of information concerning the scientific foundations of digital games, technology used to develop digital games, and the study of digital games and their design, broadly construed."[15] The DIT and two U of U faculty members presented a poster about the advantages and challenges of having The GApp Lab (chapters 8 and 12) located within a library setting.[16] The DIT discussed e-channel with the FDG attendees to gauge interest in potential collaborations.

NEW SPACES AND FUTURE LIBRARIAN ROLES

With plans to replace the current U of U School of Medicine building with four new buildings within the next five years, medical innovation spaces at the U of U Health Sciences will expand. Librarians can be embedded into these new spaces to easily partner with innovators within their working environments. This will offer a great opportunity to dedicate the IL in enhancing the U of U's rich history with medical innovation initiatives (chapter 10, table 10.1). Another opportunity for partnership is the Lassonde Studios, which opened in the fall of 2016. As mentioned in chapter 10, this new entrepreneur residential community allows students to *live*, *create*, and *launch* products and businesses. The IL, DIT, and LIT are investigating relevant ways to address the studio students' innovation-related information needs.

SUMMARY

Information transfer occurs at many stages of the innovation life cycle (chapter 2). The inclusion of the DIT and IL within innovative spaces enables them to easily identify information needs and develop ways to effectively meet those needs. This is a tremendous benefit to the U of U Innovation Ecosystem and other innovation professional associations. Being able to collide and share hallway conversations on a daily basis has offered the CMI's ADE and executive director, The GApp Lab director, the EHSL director/DIT, and the IL opportunities for collaboration that would have otherwise gone unaddressed. Participating in professional innovation association conferences has also provided a comprehensive perspective of the information needs of innovators. Librarians and innovators are natural partners and jointly can accelerate innovations in university settings. Refer to chapter 16 to read about the DIT's vision of the natural combination of information and innovation.

NOTES

1. James P. Womack, Daniel T. Jones, and Daniel Roos, *The Machine That Changed the World: The Story of Lean Production—Toyota's Secret Weapon in the Global Car Wars That Is Now Revolutionizing World Industry* (New York: Free Press, 1990).

2. Tallie Casucci, "LITE Awards $2000 to B2B Innovation Team," eSynapse: Newsletter of the Spencer S. Eccles Health Sciences Library, 2015, http://library.med.utah.edu/esynapse/2015-april-june-vol-30/lite-awards-1000-to-b2b-innovation-team (accessed June 3, 2016).

3. Tallie Casucci, "Library Innovation Team Award Winner at Bench-2-Bedside," eSynapse: Newsletter of the Spencer S. Eccles Health Sciences Library, 2016, http://library.med.utah.edu/esynapse/2016-april-june-vol-31/library-innovation-team-award-winner-at-bench-2-bedside (accessed June 13, 2016).

4. National Science Foundation, "NSF Innovation Corps (I-Corps™)," 2016, https://www.nsf.gov/news/special_reports/i-corps (accessed May 22, 2016).

5. Alexander Osterwalder and Yves Pigneur, *Business Model Generation: A Handbook for Visionaries, Game Changers, and Challengers* (Hoboken, NJ: Wiley, 2010).

6. University of Utah Libraries, "Innovation: Welcome," 2016, http://campusguides.lib.utah.edu/innovate (accessed June 3, 2016).

7. Tallie Casucci, Erin N. Wimmer, and Jean P. Shipman, "Business Model Canvas Meets Evidence: The Intersection of Innovation Tools" (paper presented at the VentureWell Open 2016 Conference, Portland, OR, March 4, 2016).

8. Department of Biomedical Engineering, University of Utah, "Welcome to Utah BioInnovate," 2016, http://www.bioinnovate.utah.edu (accessed May 22, 2016).

9. Jean P. Shipman, Roger A. Altizer, Jose Zagal, Tallie Casucci, and Tina Kalinger, "Librarians as Matchmakers: Using Dating Sites as a Model for Collaboration" (paper presented at the annual meeting of the Medical Library Association, Austin, TX, May 18, 2015).

10. Erica Lake and Jean P. Shipman, "Doodle Health: Developing Pictographs for Minority Health through Gaming" (paper presented at the annual meeting of the Medical Library Association, Austin, TX, May 19, 2015).

11. Digital Games Research Association, "About Us," 2016, http://www.digra.org/the-association/about-us (accessed June 3, 2016).

12. VentureWell, "History," 2016, https://venturewell.org/history (accessed June 13, 2016).

13. John Langell, Patrick D. Loftus, Craig T. Elder, et al., "Creating a Benchmark Medical Technology Entrepreneurship Competition: The University of Utah Bench-to-Bedside Medical Device Design Competition" (paper presented at the National Collegiate Inventors and Innovators Alliance Open 2014 Conference, San Jose, CA, March 21–22, 2014); Tallie Casucci, Jean P. Shipman, John T. Langell, and Megan McIntyre, "Opening New Doors: Unique Innovation Partners Unite" (poster presented at the VentureWell Open 2015 Conference, Washington, DC, March 20–21, 2015); Patrick D. Loftus, Craig T. Elder, David Morrison, et al., "Improving the Quality and Number of Intellectual Property Patents Filed by Student Entrepreneurs" (poster presented at the VentureWell Open 2015 Conference, Washington, DC, March 20–21, 2015); Jean P. Shipman, Tallie Casucci, Christy Jarvis, Nancy Lombardo, Melissa Rethlefsen, and Jeff Folsom, "e-channel: An Innovation Dissemination Venue" (poster presented at the VentureWell Open 2015 Conference, Washington, DC, March 20–21, 2015); Tallie Casucci, "The Art of Matchmaking: Connecting Teams to the Right Mentor" (paper presented at the VentureWell Open 2016 Conference, Portland, OR, March 4, 2016); Tallie Casucci, Erin N. Wimmer, and Jean P. Shipman, "Business Model Canvas Meets Evidence: The Intersection of Innovation Tools" (paper presented at the VentureWell Open 2016 Conference, Portland, OR, March 4, 2016); Jean P. Shipman, Christy Jarvis, and Chad Johnson, "Calling All Failures: Contribute to e-channel" (poster presented at the VentureWell Open 2016 Conference, Portland, OR, March 4, 2016); Barbara A. Ulmer, Christy Jarvis, Jean P. Shipman, and Andrew Maxwell, "Video Vault: Advice at your Fingertips" (poster presented at the VentureWell Open 2016 Conference, Portland, OR, March 4, 2016).

14. Tallie Casucci, Jean P. Shipman, Roger A. Altizer, and John T. Langell, "Revolutionizing Game Creation Partners: Health Care Professionals, Including Librarians, and Game Scholars Unite" (paper presented at the annual conference of the Digital Games Research Association, Snowbird, UT, August 4, 2014); Tallie Casucci, Jean P. Shipman, and Roger A. Altizer, "Unravelling Medical Game Research: Informing Players of Foundational Evidence" (paper presented at the annual conference of the Digital Games Research Association, Snowbird, UT, August 5, 2014).

15. Foundations of Digital Games, "Foundations of Digital Games," 2016, http://www.foundationsofdigitalgames.org (accessed June 4, 2016).

16. Tallie Casucci, Jean P. Shipman, Roger A. Altizer, and Jose P. Zagal, "Shhh! We're Making Games in the Library and You Can Too" (poster presented at the First International Joint Conference of DiGRA and FDG, Dundee, Scotland, August 1–5, 2016).

Chapter Twelve

Applying Innovation to Patient Education and Behavior

Roger Altizer Jr. and José Zagal

OVERVIEW

Games and apps are cool. There is no way around it. For the next five years (and possibly longer), people will continue to be interested in *mHealth* and digital medicine innovation—and for good reason. Good games and apps are systems that engage people voluntarily while also requiring learning and breeding creativity. They are software you learn to use through discovery, not only direct instruction. People enjoy games and apps even though playing them may be a lot of work. Users spend massive amounts of time using smartphone software ranging from Facebook to Angry Birds. As a result, it is no wonder that the study and creation of health-based apps and games is growing.[1]

The popularity of *mHealth* (games and apps), the increased study of them, and the potential impact they have is what motivated collaboration between University of Utah (U of U) Health Sciences and game developers to create a lab that would generate patient- and clinician-facing software. The initial creation of the Therapeutic Games and Apps Lab (The GApp Lab) is discussed at length in chapter 8. This chapter focuses on the lab's processes and pipelines used to develop innovative and novel health-related software and provides examples of its patient-facing *mHealth* games and apps.

THE GAPP LAB

At this point in its evolution, the lab is staffed with a director; several faculty collaborators from Health Sciences and the Entertainment Arts and Engineering (EAE) Program; two full-time project managers; 20 to 35 funded EAE graduate students working as artists, engineers, and producers; and a dedicated innovation librarian (chapter 9), with assistance from the Spencer S. Eccles Health Sciences Library (EHSL) director as needed.

How Faculty Interact with The GApp Lab

Across the U of U, faculty are often interested in and encouraged to create patient-facing software as part of their research agendas. It is not uncommon to see the National Institutes of Health and the National Science Foundation (NSF) encourage innovative software in requests for proposals, and universities themselves encourage and celebrate the creation of patient-facing software. Some of these resulting systems are large. For example, the U of U has put tremendous effort into and derived great results from Value Driven Outcomes, a system to improve quality of care by decreasing costs and increasing value.[2] Other systems are smaller. For example, *Health X* is a game designed to detect and treat lazy-eye disorder. It began as a student-driven class project that was further developed by one of the original team members. It has won several serious games contests, received an I-Corps NSF seed grant, and is currently being supported by the Moran Eye Center (http://healthcare.utah.edu/ moran) while it undergoes clinical trials.[3] These kinds of efforts, at multiple levels of complexity, funding, and resources, are part of a growing ecosystem that encourages the creation of patient-centered software. Furthermore, the recent popularity of start-up culture is encouraging many people to see games and apps for patients as a space for cultivating new businesses. The health app ecosystem is not going away soon, nor is the enthusiasm for it, but it is a complicated field to navigate for uninitiated faculty.[4]

The GApp Lab runs between 7 and 10 funded research projects every semester. These projects may be supported by grants that faculty within or outside the lab have been awarded, sponsored by private companies or foundations, or funded internally by U of U Health Care and academic programs. Internal funding is often provided to encourage collaboration on innovations or to develop solutions to challenges. Each project has one or more EAE faculty, one project manager from The GApp Lab to oversee the project's completion, and three to six graduate students who do the majority of the software development (including design, programming, art, etc.). Some projects are completed in a single semester, while others take several years.

Most of The GApp Lab's projects arrive via referral. For example, a company asks their contacts at the U of U if they know of anyone who can help them determine if stress can be detected via a wearable fitness tracker, and it is suggested that they contact The GApp Lab. Or, as another example, a professor of dance who wants to create an app exploring how choreographic thinking may help autistic children is told by others to reach out to The GApp Lab.

These referrals occur for several reasons. First, the EAE Program is one of the top-ranked game development programs worldwide; it has been ranked number one by the *Princeton Review* for several years. Second, because of the number of projects and partners The GApp Lab has achieved, it has quickly gained a local reputation as a place to ideate and develop projects, attracting faculty who want to apply for grants or need help fulfilling awarded ones. Finally, unlike traditional collaborations that tend to be ad hoc, The GApp Lab has developed a system and process for onboarding and developing projects. When anyone with an idea approaches the EAE Program, the Center for Medical Innovation (CMI), or the EHSL, they know they can set a meeting with either the director of, or a project manager from, The GApp Lab who will help them unpack their project.

TYPES OF PATIENT-CENTERED SOFTWARE CREATED AT THE GAPP LAB

Clinical, laboratory, and patient-facing software is created by The GApp Lab. Three of the lab's patient-centered games and apps are featured in this chapter. It is important to note that the physical location of The GApp Lab within the EHSL facilitated being able to easily meet with Health Sciences partners as well as immediate access to the innovation librarian, the EHSL director, and other library resources, which were key in the development of the games and apps described.

Behavior Change: Em(B)r *MediGarden*

MediGarden is an electronic medical record (EMR)–linked gamified app intended to increase compliance for patients who take prescription medications. It is a pill minder that synchronizes with patients' medical records, lists their prescribed medications and administration instructions, and gives pictures of the medications (for easy identification). *MediGarden* also allows for customization. Patients can set alarms and reminders to take their medications, remove medications from the list of prescriptions needing reminders (without deleting the prescriptions), and change the look and nicknames of their medications.

MediGarden illustrates two major innovations. First, it is a dynamic EMR-linked pill minder, meaning that patients can see not only their current prescribed medications but also subsequent changes clinicians make to their treatment plan. Second, it is gamified. As patients interact with the app and take their medications as prescribed, they are given virtual plants to water and manage. The more they stick to their prescribed medication plans, the more they can tend and grow their gardens. There is also a small built-in luck-based component. As patients interact with *MediGarden*, they can spin a wheel to win prizes, such as new plants, watering cans, and the like; the more they interact with the app, the more chances they have of winning such prizes.

MediGarden can be applied in any number of ways. It can be used as a simple list of medications, as a pill minder, or as a gamified motivational tool. Patients choose which features are important to them as well as how much they wish to interface and in what ways they wish to engage. *MediGarden*'s goal is to motivate, not frustrate.

MediGarden is the first of multiple games and apps designed by The GApp Lab that link with an EMR system for patient use. The name Em(B)r stands for *building for the EMR*. Seed funding for this project was provided by internal funds to meet the desire of the health system in the state of Utah to create a pipeline for EMR-enabled apps. Medication compliance was identified as both a major need and a good test case for the use of SMART on FHIR[5] (an open standard for EMR integration) to develop games and apps.

Simulations and Trainers: *Tetra-Ski* and *Tetra-Sail*

Tetra-Ski and *Tetra-Sail* are two projects that are part of the Tetra-Controller initiative. The medical director of the Spinal Cord Injury Acute Rehabilitation program at the U of U asked The GApp Lab to apply for a Craig H. Neilsen Foundation (http://chnfoundation.org) grant with him and his TRAILS organization (http://healthcare.utah.edu/rehab/support-services/trails.php). The medical director was looking to extend the work he was doing on adaptive sports equipment. For perspective, Utah is one of the outdoor recreation capitals of the world; helping spinal-cord-injury patients participate in outdoor activities is seen as an important part of both rehabilitation and wellness efforts. The foundation grant would support the creation of a simulator to accompany an existing adaptive sailboat for spinal-cord-injury patients, and the development of an adaptive ski chair. Additionally, a third component to be designed was an innovative solution to encourage wheelchair users to perform *pressure reliefs* (an exercise designed to reduce pressure ulcers in spinal-cord-injury patients).

Initially, The GApp Lab task was well defined: to create software to simulate the experience of using the *Tetra-Ski* and *Tetra-Sail* devices. In initial meetings, several *Design Box* sessions were conducted (described later in this chapter) to scope the project. After several meetings, it was decided that the simulators should be game simulations (meaning they provide a believable representation of the experience as opposed to being an accurate representation of reality), as it would be faster to create the simulations using a game engine, in this case *Unity*, and game simulation would better meet the initial training goal. Both realistic scenarios and abstract ones were created in order to enhance teaching the desired skills (in this case, how to use the adaptive sailboat or ski chair for *Tetra-Sail* and *Tetra-Ski*, respectively). For example, while users of the Tetra-Controller ski chair or sailboat may not ever go through a slalom, practicing slalom in the simulator could be useful for skill development. Similarly, while it may not be desirable to have users of the physical ski chair and sailboat jump, crash, or sink, it was both motivational and educational to allow them to do so in the game. For this project, it was better to have a believable and enjoyable skiing and sailing experience as opposed to a realistic simulation of one.

This development experience emphasized vocabulary differences between game developers and medical professionals that required special attention to articulate these knowledge representation differences. For example, *realistic* driving games on computers and home game consoles *cheat* by using *fake* physics to provide players with an enjoyable driving experience; the focus is on believability over realism. *Cheat* and *fake* are standard game development terms meaning to bend the rules of physics to enhance the experience, but when used with rehabilitation faculty, they interpreted the terms to mean that the game developers were trying to cut corners or do shoddy work.

These kinds of interdisciplinary collaborations are also an opportunity to cross-pollinate innovations. It is common for games to have customizable controls so that players can find a control scheme that is comfortable, familiar, or advantageous in some way to them. For example, someone may prefer to use the Shift key on a keyboard to make a character jump instead of the space bar. From the TRAILS team, gamers learned that many of the devices designed for patients with disabilities are not customizable. Thus, while some people may be able to move their arms in one direction more easily than another, most powered wheelchairs and other adaptive devices do not allow for much (if any) customization of the controls. This discovery led to the Tetra Universal Controller.

The Tetra Universal Controller allows users to customize joystick controls as well as *sip* and *puff* controls that use air pressure by sipping/inhaling and puffing/exhaling on a straw. For example, if a user wants to turn the ski chair left with one tap and use two taps to turn the chair right, they can customize it that way. If they want to set a long *puff* of air to unfurl the sail on the sailboat and two *puffs* and a *sip* to put the motor in reverse, they can also set the controller to do that. Thus, the developed Tetra Controller games became skill development and testing devices and offered users the ability to customize the games to suit personal needs and have those customizations transfer to physical adaptive sports equipment.

Educational Tools: *Arches Saves Your Bacon*

Arches Saves Your Bacon is a mobile Web game that was funded by the now-defunct Arches Health Insurance Company. The company first approached the CMI to create something to help young adults understand the newly implemented health care marketplace resulting from the Affordable Care Act. The CMI introduced *Arches* to The GApp Lab to create a solvable problem statement with this topic in mind. After multiple *Design Box* sessions, all agreed that it would be advantageous to create a game to help college students understand that paying for health insurance plans would actually be cheaper than having to pay actual health care costs.

The development of this game required more collaboration with the EHSL innovation librarian than any other project to date. The game included a *press-your-luck* mechanic (much like dated television game shows), where students could choose six activities, with the game indicating if they would get hurt by engaging in those activities and the associated health care costs of the resulting injuries. It was important to be realistic, as the goal was to encourage college students to weigh the pros and cons of having health insurance. To obtain these costs, the innovation librarian became a member of the development team, providing members with data on the incidence of accidents and their affiliated costs.

Additional Creations

These three game examples offer only a small glimpse into the variety of games and apps developed by The GApp Lab. In its three years of existence, more than 40 games and apps have been created, not all of them consumer facing. Other kinds of tools have been developed, such as a website to help researchers find collaborators, a plug-in for clinicians to view lab reports in an interactive manner, and games and interactive infographics for a variety of research studies. In all cases, these projects were developed by some combination of students, faculty (including librarians), clinicians, and researchers as well as industry and community partners.

DESIGN WORK NAVIGATION

As previously described, there is need for faculty to engage in ideation and design for their software products. At The GApp Lab, a design methodology developed by Roger Altizer Jr., known as the *Design Box*, is used to help faculty quickly engage in iterative design and focus on inductive ideation.[6] Generally, most approach the lab with a pitch: "I have a great idea! Would it not be cool if we built a game where you did X." The ideas are really solutions. Whether they realize it or not, faculty pitch software as an answer.

The GApp Lab uses faculty pitches as starting points that lead to reverse engineering the problems to be solved and the further refinement of the faculty's ideas.

The *Design Box* is an inductive participatory design practice that was originally inspired by the qualitative research method *grounded theory*,[7] the notion being that if one gets a deep enough description of a problem, the number of solutions relevant to it will seem relatively small. At the very start of a project, the research team does a *Design Box* together. Then sessions are done with end users and key stakeholders, time and budget permitting. After all of the *Design Box* sessions are held, the research team analyzes the session notes to uncover a final pitch/solution. Essentially, once the description is deep enough, one of the pitches surfaces to the top and becomes the thing that gets built.

To ensure that the *Design Box* session is successful, two brainstorming rules are enforced. First, no ideas can be shot down unless they clearly do not *fit* inside the *Design Box* (which will be clarified soon), and, second, it is best to use *yes and . . .* when addressing the ideas of others. That is, it is better to build on ideas and watch them grow and evolve rather than disagree with them.

A *Design Box* session starts with the moderator drawing a large box on a whiteboard. The four box walls are labeled audience, technology, aesthetics, and problem statement/question. In general, the first wall to get tackled is the *problem statement/question*. "What problem does this software solve?" or, put another way, "If you were to point at this software and say, 'Oh this app/game does that,' what would *that* be?" The moderator then moves on to the *audience* wall, exploring who is the end user and who are the stakeholders or gatekeepers who can help make this project happen or fail. The *technology* wall asks what technologies are available, what technologies will the end user likely have available, and whether there are any special hardware or software considerations to respect. Finally, the *aesthetic* wall explores how end users should feel while using the software. This wall tends to display lots of descriptive words, such as *hopeful* or *serious*.

At this point, the *Design Box* is, metaphorically, smaller than when the session started. The details along the four walls serve as design restrictions for the next phase: *pitching*. The moderator writes ideas inside of the box in the form of short elevator pitches. It is the moderator's job to make sure the two previously mentioned rules are enforced. While multiple ideas usually fit inside the box, one of two things frequently happen. An idea might get so much *yes, and . . .* attention that it becomes clear that this is the idea endorsed by the majority of the design team. Conversely, if no idea is immediately apparent as the way to go, then the research team examines results from the multiple *Design Box* sessions to see if one emerges as the best pitch to pursue.

DISTRIBUTION METHODS

Diffusion of innovation is a complex task for health care information technologies, including medical games and apps.[8] This is no exception for The GApp Lab. While the faculty and students are experienced developers and innovators, sharing information about the medical games and apps developed by the lab is no small task. Methods used are traditional academic venues, such as professional conferences and academic publications. The director of The GApp Lab's background in journalism has helped with getting publicity for lab project outputs. However, EHSL has taken the lead on two innovative pipelines for the diffusion of innovation: the e-channel (chapter 13) and the *U-Bar*.

e-channel

e-channel is a forward-looking dissemination platform that archives, curates, and distributes innovative digital artifacts and information. Developed in partnership with EHSL faculty and staff, The GApp Lab produced the first platform interface that offered content ranging from games and apps to videos and development documentation. This platform provides not only the distribution but also the preservation of the innovative content it stores and serves. As software production is being accepted by some academies, it is important to create venues for distributing noncommercial games and apps as well as the media that surround them.

U-Bar

The *U-Bar* is located in the EHSL's consumer health library, the Hope Fox Eccles Health Library, located in the U of U Hospital. It has iPads, a large video monitor, and student staff who assist consumers with identifying suitable *mHealth* apps. Having such a center helps with discovery, but it also serves to keep patients safe, as the recommended apps have been vetted by health sciences librarians and clinicians. The *U-Bar* offers a new venue to playtest games and apps and conduct research in addition to showcasing them in a location visited by potential users (patients and their families).

DEVELOPING GOOD CLIENT ATTRIBUTES

As more faculty are interested in developing innovative software as part of their research, it is becoming clear that many are not prepared for this kind of work. Graduate school does a good job of socializing students for future academic life. Many graduate student teaching assistants teach undergraduate courses to help them understand how to become effective teachers. They frequently receive oversight from a faculty mentor who provides feedback and guidance about their teaching skills. Students learn to write academic papers in their classes and are encouraged to submit them to various professional conferences and academic journals by their mentors. Many graduate students work in labs that provide learning opportunities, such as how to get research funding and develop research protocol strategies and, again, how to disseminate research results. In addition, there are writing workshops available to students to learn how to be successful when conducting traditional types of research.

However, outside of computer science or information technology (IT) studies, few get training in software engineering or development. This means that when faculty write grants, they include funding for someone else to develop named software. This leads to three potential problems: (1) a lack of understanding of ideation and design processes, (2) a lack of understanding of the software development cycle, and (3) a lack of understanding of how to be a client or partner when working with private software developers.

One of the key advantages of having a lab with software development capabilities located within the Health Sciences is the ability for faculty to receive help with all three identified problems. As The GApp Lab is faculty managed, the lab understands how to collaborate with faculty. It is a research lab that is focused on being partners rather than offering a client–developer relationship. The lab has the goals of doing quality research, providing opportunities for its student employees, and advancing the interests of faculty partners. Unlike a private vendor, which may be well intended but ultimately has to extract more value from a relationship than it contributes in order to be profitable, The GApp Lab does not need to be financially solvent. As it partners with other local faculty, any disputes can be handled by the university, which is interested in the success of all parties involved. Additionally, The GApp Lab is familiar not only with how faculty work but also with good development practices. Part of the value it brings to partnerships is being able to guide projects through ideation and design as well as educate partners along the way about good software development.

Finally, as part of an ecosystem supported by the EHSL, the CMI, and the EAE Program, The GApp Lab can help point partners to resources and guide projects toward success. For example, because the lab is located nearby and is an active member of the Health Sciences community, it knows what IT policies exist and can facilitate making sure that appropriate IT concerns are addressed as well as building software so that it can be supported by U of U IT. It understands that live software has to be supported, so it assists partners with maneuvering the university's legal and technical services to help projects see the light of day and not die on the vine.

Ideation and dissemination are complex issues to address. To help with this, there is an Ideation Studio within the EHSL where faculty and students can work within a space equipped with everything needed for paper prototyping and idea iteration. The *U-Bar*, mentioned earlier, also serves as a dissemination channel for completed projects, as does e-channel.

SUMMARY: THIS SEEMS LIKE A LOT OF WORK

In this chapter, interdisciplinary, innovative software development processes and facilities have been explored to see how software can emerge using a strong design methodology. While challenging, this kind of work and the process are important. Software, apps, and games are becoming increasingly pervasive in our lives, and while we may be getting better at developing them, there is still more to do. The past couple of decades have been littered with patient-facing health software that has never been used. This was mostly either because the software was poorly designed and thus hard to use or because it was ineffective in terms of expected outcomes. These two issues can be solved by engaging in collaborations like the ones described.

Advances in game design and interfaces enable usable, interesting, and conducive user experiences. These experiences need to be combined with the increasing knowledge of how to best help patients solve and manage their personal health issues. Furthermore, recognition of the fact that software tools do not exist in isolation of the sociotechnical systems in which they reside is required. The future of health sciences will not be one of discrete software used in isolation; rather, it will be one of software used in the context of broader and deeper connected systems. Understanding how to best leverage and integrate these multiple systems mandates interdisciplinary collaborations and the application of diverse expertise. The scope of EMR software is growing; for example, it is now possible to build (1) wearable fitness trackers that directly write collected data to patients' EMRs, (2) virtual reality systems that allow patients to explore complicated health problems in an immersive and embodied way, and (3) educational apps that help patients engage and comply with their treatment plans. While the future will host even more *mHealth* software, it is important to create this software in such a way that makes people want to use and readily adopt it.

In order to achieve this future, especially in the context of a university, interdisciplinary innovation labs, like The GApp Lab, are crucial, as they provide a way to leverage the unique resources of different groups. In the case described, these groups are the key partners of the EHSL, the CMI, and the EAE Program. It is only through their combined expertise and cooperation that the common pitfalls of making patient-facing software are alleviated.

NOTES

1. Christophe Gaudet-Blavignac and Antoine Geissbuhler, "Serious Games in Health Care: A Survey," *Yearbook of Medical Informatics* 7 (2012): 30–33. https://www.ncbi.nlm.nih.gov/pubmed/22890338 (accessed November 10, 2016); Guido Giunti, Analia Baum, Diego Giunta, et al., "Serious Games: A Concise Overview on What They Are and Their Potential Applications to Healthcare," *Studies in Health Technology and Informatics* 216 (2015): 386–90, https://www.ncbi.nlm.nih.gov/pubmed/26262077 (accessed November 10, 2016).

2. Kensaku Kawamoto, Cary J. Martin, Kip Williams, et al., "Value Driven Outcomes (VDO): A Pragmatic, Modular, and Extensible Software Framework for Understanding and Improving Health Care Costs and Outcomes," *Journal of the American Medical Informatics Association* 22, no. 1 (2015): 223–35, doi:10.1136/amiajnl-2013-002511.

3. Daphne Chen, "U. Students Invent Video Game to Help Diagnose Lazy Eye in Kids," posted August 28, 2016, https://www.ksl.com/?sid=41255513&nid=148 (accessed November 10, 2016).

4. Jessice L. Baldwin, Hardeep Singh, Dean F. Sittig, and Traber Davis Giardina, "Patient Portals and Health Apps: Pitfalls, Promises, and What One Might Learn from the Other," *Healthcare*, October 3, 2016, doi:10.1016/j.hjdsi.2016.08.004.

5. Joshua C. Mandel, David A. Kreda, Kenneth D. Mandl, Isaac S. Kohane, and Rachel B. Ramoni, "SMART on FHIR: A Standards-Based, Interoperable Apps Platform for Electronic Health Records," *Journal of the American Medical Informatics Association* 23, no. 5 (2016): 899–908, doi:10.1093/jamia/ocv189.

6. Roger Altizer and José P. Zagal, "Designing Inside the Box or Pitching Practices in Industry and Education," *Proceedings of the 2014 DiGRA International Conference*, August 3–6, 2014, Snowbird, UT, http://www.digra.org/digital-library/publications/designing-inside-the-box-or-pitching-practices-in-industry-and-education (accessed November 10, 2016).

7. Clay Spinuzzi, "The Methodology of Participatory Design," *Technical Communication* 52, no. 2 (2005): 163–74, https://repositories.lib.utexas.edu/bitstream/handle/2152/28277/SpinuzziTheMethodologyOfParticipatoryDesign.pdf?sequence=2 (accessed November 10, 2016); Barney G. Glaser and Anselm L. Strauss, *The Discovery of Grounded Theory: Strategies for Qualitative Research* (Chicago: Aldine, 1967).

8. Anthony G. Bower, *The Diffusion and Value of Healthcare Information Technology* (Santa Monica, CA: RAND Corporation, 2005), http://www.rand.org/content/dam/rand/pubs/monographs/2006/RAND_MG272-1.pdf (accessed November 10, 2016).

Chapter Thirteen

e-channel

A Platform for Disseminating Innovators' Outputs

Christy Jarvis, Chad L. Johnson, and Jean P. Shipman

Faculty tenure and promotion decisions have long been inextricably tied to scholarly publishing metrics. In evaluating researchers' contributions to their fields, academic review committees, funding agencies, and institution leaders have come to rely on various article citation analyses that attempt to quantify individuals' productivity, visibility, reputation, and impact. In addition to influencing tenure and promotion, scholars' publishing profiles are likely to affect the way in which their work is funded, circulated, evaluated, and archived. It is not surprising, then, that savvy researchers carefully choose what and where they want to publish, basing their decisions on a host of factors, including the venue's reputation, audience reach, and publication policies, such as the peer review process and speed of publication.

With citation metrics serving as the gold standard for evaluating research quality and significance, scholars in emerging fields, such as medical device design and therapeutic health applications, find themselves disadvantaged. Only a small portion of their work leads to the type of results commonly reported in traditionally published journals. This leaves vast quantities of hard-earned knowledge hidden from view—unpublished, unrecognized, and inaccessible to others. This lack of a dissemination venue for creative scholarly output prevents faculty members at academic institutions from receiving credit for their contributions and impedes progress by depriving future generations of innovators from building on the important work of their predecessors.

To address this imbalance in publishing opportunities for creative and innovative works, the Spencer S. Eccles Health Sciences Library (EHSL) at the University of Utah (U of U) partnered with the Center for Medical Innovation's (CMI's) Therapeutic Games and Apps Lab (The GApp Lab) and other institutional stakeholders to develop e-channel, an online multimedia platform that collects and publicizes the work of innovators across many disciplines but particularly the health sciences. This chapter describes the evaluation process that led to the EHSL's decision to create e-channel, discusses the pitfalls and challenges encountered throughout its development, and offers guidance for other health sciences librarians interested in establishing similar partnerships to promote nontraditional innovative scholarly output at their institutions.

BACKGROUND

The EHSL at the U of U is one of three university libraries that serves the Colleges of Health, Nursing, and Pharmacy and the Schools of Dentistry and Medicine as well as the U of U Health Care Hospital and Clinics. It also is the library for many interprofessional centers and institutes. The mission of the EHSL is "to advance and transform education, research, and health care through dynamic technologies, evidence application, and collaborative partnerships. The library contributes to the success of health professionals, students, researchers, innovators and the community."[1]

Through a series of construction projects and the removal of a majority of the EHSL's print collection,[2] the resulting large footprint of space enabled the installment of several U of U innovation units (chapter 3). With fewer traditional library roles to perform (e.g., a limited number of books and journals to circulate and reshelve), EHSL staff were freed to focus on new opportunities to use their skills to benefit the U of U, including those that were of value to the new innovation units. Being in close proximity with innovators encouraged conversations about innovators' information needs, and one that surfaced repeatedly was the need to have an outlet for their creative output that is recognized by the U of U as being academically appropriate for faculty promotion, tenure, and retention recognition. Most innovators do not obtain extramural funding grants, such as R01s, nor do they publish many journal articles about their innovations, such as medical devices or therapeutic games and apps. Discussions with gamers from The GApp Lab revealed that creative content was not being preserved (e.g., failed games, unfinished games, and earlier game versions) due to a lack of infrastructure and platform for capturing this content. Even completed game components were being lost, as graduating students were taking with them the games and intellectual property created while at the U of U. These various innovators needed a formal peer-recognized outlet for their creative and scholarly contributions.

A consulting firm, Delta Think, Inc., assisted the EHSL with learning more about how it could address this identified scholarly communication need. Consultants met with many stakeholders from across the U of U and with a smaller joint EHSL and innovator faculty team, including the CMI executive director. Through a series of focus groups and personal interviews, the consulting firm issued a report indicating that a new platform for innovation might be one means of addressing the distribution void, a platform that would publish nontraditional, multimedia kinds of material. Based on that report, e-channel was developed in the spring of 2014 and launched in early 2015.

e-channel is an interactive platform designed to capture and disseminate the creative output of innovators in all disciplines but particularly the health sciences. This eclectic hub offers a venue for innovators and researchers to share their results, receive recognition, and contribute to their scholarly disciplines while also ensuring that others can build on the work reflected. This multimedia portal enables the transfer of innovative ideas regardless of format and at various stages of creation, implementation, or even failure. The name e-channel reflects the eclectic, electronic, and entrepreneurial spirit of the platform while also highlighting the creator's name, the EHSL. The word *channel* is a familiar word among innovators, as one step of the innovative process is to determine distribution channels for generated products and outputs.

METHODS/PROJECT TIMELINE

In the spring of 2014, with the concept for e-channel in place, EHSL staff turned their attention to making the platform a reality. A core EHSL team, consisting of five faculty and one staff member, was created to oversee the process of bringing e-channel to life. The team's first task was to pursue a collaborative partnership with two of The GApp Lab designers to create a logo that would convey e-channel's place as the home for nontraditional creative scholarly output at the U of U and beyond. Several ideas and versions were reviewed, debated, and modified, but by August 2014, the EHSL e-channel team had settled on a logo.

While the logo design discussions were under way, the EHSL also drafted and submitted a proposal for e-channel site design work. The signed contract between the EHSL and The GApp Lab called for 1,400 hours of e-channel site design work from a student team to be completed during the fall 2014 semester. The student team would be comprised of an engineer, an artist/

designer, a producer, and a research assistant working under the direction of a faculty mentor.

Throughout the fall semester, the EHSL met frequently with The GApp Lab student team to review design concepts and evaluate e-channel platform iterations. Several options were considered.

Iteration 1. Digital Asset Management Database Styles: Using ContentDM

This was the initial plan used for creating the e-channel site. It relied on an underlying database called ContentDM, a digital asset management tool used by the EHSL to catalog and preserve documents. The original idea was to create a landing page for the website home page and then to restyle the database pages to match. However, on detailed investigation of this option, it was determined to be out of scope for The GApp Lab team and the amount of time allotted for the project. It was also limited by the access to the Digital Asset Management database server to accomplish the restyle of its pages. Since the project was still in its early stages and EHSL wanted a unique and eye-catching landing page to help market the product to potential clients to help grow the site, it was decided to scope the project down to a flashy landing page.

Iteration 2. Parallax Scrolling

The second iteration was a version of a currently popular style of websites called parallax. This involves the background moving at a different rate than the foreground as the user scrolls. The GApp Lab created a test version of this, but it didn't fit the EHSL's needs due to the inability to create subcategories of content.

Iteration 3. Animated Mind Map Bubbles

This was the version that The GApp Lab and EHSL teams agreed to develop fully. It followed the design pattern of a mind map that helps to organize data. It consisted of a central bubble with the category or subject term inside with lines extending out to other bubbles that related to that subject.

Inside of each bubble was a title of the program and a watermark background logo. These bubbles linked outside to a landing page for that program. All the data for these pages were generated dynamically from a JSON file acting as a small database. EHSL was able to update or change this document and have it load right into the updated website. This allowed the EHSL to add new bubbles as content from new departments, associations, or student competitions was collected and disseminated. The GApp Lab accomplished this interface using the HTML 5 canvas element (see figure 13.1).

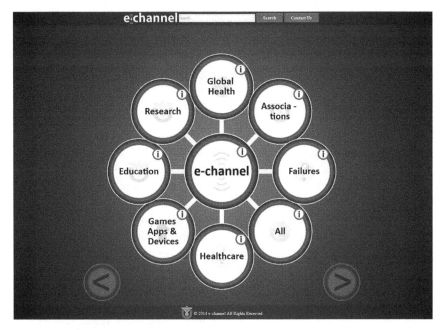

Figure 13.1. Original e-channel Interface.

By late December 2014, The GApp Lab student design team handed over the completed site to the EHSL. Shortly thereafter, the EHSL hired a full-time e-channel content specialist to manage the work flow and technical demands of the site. This includes tasks such as developing Web pages for e-channel collections as they emerge; uploading e-channel objects to various content management systems, such as YouTube, Kaltura, ContentDM, and WordPress.com; and applying basic metadata to e-channel items and collections.

GUIDING PRINCIPLES

Meanwhile, the rest of the EHSL e-channel team was at work formalizing e-channel's policies, submission standards, and collection development goals. These internal guidelines ensure a clear understanding of e-channel's mission and differentiates the platform from other existing U of U efforts, such as the institutional repository. These guiding principles are outlined below.

Philosophy

The e-channel team will create, gather, organize, maintain, disseminate, promote, and preserve digital content that reflects the process and results of innovative projects in all health science disciplines.

e-channel will collect content that can be publicly shared. In some cases, access may need to be restricted in order to adhere to copyright law or satisfy privacy concerns. In other cases, new content may be collected or existing content enhanced to build fee-based collections that are not freely shared.

In-Scope Content

The focus of e-channel is on unique content produced in conjunction with innovative projects and ideas. This includes slides, figures, recordings, photographs, templates, abstracts, summaries, posters, nonpublished manuscripts, or other works that purport to convey the essence of an individual, group, or institution's innovation.

e-channel accepts content that captures the innovation process at all stages, including idea generation, idea execution or implementation, and postanalysis. Submissions are encouraged that relate to the myriad avenues for innovative expression, including process improvements, innovative approaches to research, teaching, health care, new technologies, brainstormed ideas, created health applications, games, videos, medical devices, and so on.

e-channel collects representations of innovative work regardless of the outcome. The collection seeks to capture both successes and failures. Particular emphasis is placed on capturing items and sustaining collections that are of *enduring value* rather than of temporary significance.

Out-of-Scope Content

e-channel is not meant to be solely a digital archive and does not collect works for the sake of preservation only. Examples of out-of-scope works include published articles, theses, and dissertations, data sets, and conference posters and proceedings (if not related to the innovation theme) and lectures, talks, panel discussions, or webinars that do not satisfy the requirement of representing an innovative concept or innovative means of dissemination.

e-channel also does not collect works in file formats requiring specialized software to render, such as GIS files.

Audience

With a stated purpose of disseminating included works as widely as possible, e-channel content is made freely and openly available when possible. Specific emphasis is placed on reaching current and future generations of students, researchers, and innovators so they can access and build on the work of others. It is not limited to content produced by U of U individuals.

Collaboration

The development and management of e-channel involves project management and strategic planning; metadata creation and management; the development and use of systems to create, manage, preserve, and deliver digital content; and the ongoing assessment of collections and services. The success of these activities hinges on collaboration. The e-channel team works closely with other university departments, editorial boards, and external partners as appropriate. Where appropriate and with permission, e-channel will maintain links to content hosted by other organizations, although the preference is for content to be stored locally as part of e-channel.

Ownership

Unless otherwise negotiated, the individual or organization submitting content for e-channel publication shall retain copyright in and to its work, and e-channel shall provide proper attribution in the exercise of its role as collector, publisher, and disseminator of the work. The e-channel team reserves the right to apply logos or other methods of branding to identify content as part of e-channel.

Relationship to Other Collections

Materials that build on existing digital collections, whether at the U of U or elsewhere, can improve research in specific subject areas and increase the value of the materials themselves. Potential for collaborative collection building is an important factor to consider.

Maintenance and Removal of Content

The EHSL is not interested in assuming copyright of any submitted content, so it created a publishing agreement form to be signed by the contributor at the time of content submission. The agreement clearly indicates that the contributor retains copyright ownership, but it grants the EHSL the right to publish, reproduce, digitize, and preserve the work via the e-channel platform.

This policy is designed in part to ensure e-channel's development is of high quality, useful and usable, and cohesive. It is possible, however, that individual objects or entire collections may need to be removed or deaccessioned for reasons of weeding, storage, or copyright dispute, among others. These decisions regarding disposition of content are made in conjunction with the content contributor(s) and the EHSL e-channel team.

The EHSL e-channel team focuses on capturing elements of innovative projects that are of use to future generations of innovators and entrepreneurs.

Discussions with thought leaders and experts in the field helped the EHSL to identify information related to failed endeavors that would be most beneficial to others. These data points serve as the foundation for a structured template created by the EHSL to standardize the way in which this crucial information is captured, organized, and presented to users (http://library.med.utah.edu/e-channel/e-channel-failure-form).

With these internal operational elements in place, the EHSL team began work in earnest to identify suitable content to make available via e-channel. Partners and content contributors from within the U of U included U of U Health Care, the CMI and its The GApp Lab, the Entertainment Arts and Engineering Program, the Evidence-Based Practice Council, and individual faculty members. Sample programs are briefly described below to give the reader an idea of the content types and formats included:

- U of U Health Care—a library of A3 Lean project templates containing suggested process improvements and efficiencies for health care organizations
- Bench to Bedside—videos and reports on the outcome of the yearly competition that introduces medical students, engineering students, and business students to the complex cycle of medical device innovation (chapter 10)
- Healthi4U—videos produced by students as part of a health education annual competition
- The GApp Lab—multiformat *wrap kits* consisting of various game design elements, such as artwork, style guides, and demo videos
- Faculty projects—a neuroanatomy video lab showing brain dissections and a video essay series encouraging health care reform at pediatric academic institutions

Partners and content contributors from outside the U of U include VentureWell (formerly NCIIA), the Digital Games Research Association (DiGRA), and the Canadian Innovation Centre. Their programs are outlined below to give an idea of the content types and formats disseminated via e-channel:

- VentureWell—proceedings from the organization's annual Open Conference, including papers, posters, and presentations.
- DiGRA—annual conference proceedings, including papers and posters.
- Canadian Innovation Centre—*The Innovation Vault*, an organized collection of videos from a wide variety of industries centered on the topic of innovation. *The Innovation Vault* contains lectures, formal classes, orations, and interviews from experts from both the private sector and academia (chapter 15).

OUTCOMES/RESULTS

By the end of 2015, e-channel was home to 17 distinct programs that showcased creative scholarly output in health care, education, research, and game or app development. As the site progressed and new content was collected, the EHSL e-channel team encountered two obstacles that threatened to limit the platform's growth. The first concerned the number of programs that could be displayed using the *bubble* navigational structure on the site's home page. The design accommodated a maximum of eight bubbles, meaning that users had to click through additional pages to see any collections that were not shown on the landing page. The second impediment to the platform's growth stemmed from the site's infrastructure, which made it necessary to design and create new separate landing pages for each program. This was time consuming for EHSL staff and made searching across the various e-channel programs impossible. This lack of cross-program searching limited the site's usability, as content was discoverable only by browsing—meaning that users had to know *where* the content of interest was likely located. After one year of development of the original e-channel site, EHSL staff determined that the mind map design did not meet the requirements of its growing programs and began exploring alternatives. In early 2016, the e-channel team decided to migrate existing disparate programs into a single WordPress.com theme that supports cross-program searching and multiformat discoverability. In addition, the WordPress.com platform provides e-commerce functionality that supports EHSL's future plans for offering paid access to specific programs as a way of generating revenue to make the platform self-supporting.

DISCUSSION

While the need for a distribution channel such as e-channel was expressed by many innovators, the U of U Innovative Ecosystem has not yet adopted e-channel as the sole holder of record of their collective innovative output. Members of the Ecosystem are fine with content being duplicated and hosted by e-channel, but they wish to retain their generated unique content on their respective websites as well. The dual location of material decreases the U of U's dependency on e-channel and thus its usage.

Working with innovators who thrive on competition and on developing proprietary output, potentially patentable and thus profitable, has been a terrific opportunity to observe different professional cultures. Librarians tend to be very open with their content and output, whereas innovators are not as willing to share. Several development ideas for e-channel have been curtailed as a result. A great example of this is with mapping intellectual processes and resulting successes and failures. Innovators do a lot of pivoting of their ideas along the development process. Capturing this intellectual flow and diversion of ideas through templates was something the creators of e-channel wanted the Bench to Bedside student teams to do. It soon became evident that the teams were not comfortable with this type of documentation and were not willing to expend the time and effort to record their processes.

The idea of collecting failure stories was also something that has not been fruitful. Everyone wants to hear about others' failures and believes there are rich lessons to be learned from such reports, but only one reported failure has been submitted to e-channel to date. Is *failure* too harsh of a word for public exposure? Are innovators not willing to share their failures, as today's failure, with a few tweaks or with time, may become tomorrow's success? Sharing information may prevent innovators from capitalizing on their intellectual property. Timing is indeed everything in innovation, and what may not work today might, with technological advances, be a prime product of tomorrow. In addition, something introduced by an innovator when a need does not exist may, again with time, reemerge when it is needed and result in that failure becoming a profitable success.

Another lesson the e-channel creators are learning is deciding what content to collect, how to collect desirable content, and preventing scope creep. What does *innovation* really mean, and how broadly should it be interpreted? As e-channel is still in its development stage, content has been gathered as individuals, groups, and associations have been willing to contribute it. Innovation, for e-channel purposes, is being defined as product innovation, process innovation, business innovation, service innovation, and organizational innovation. It also includes innovations in health care, education, research, global health, and technology.

Creating such an entity as e-channel has provided EHSL faculty with the opportunity to report about its development process and existence, including the successes, challenges, and failures experienced. Posters and presentations have been given at the Medical Library Association (MLA) annual meetings,[3] the Midcontinental Chapter of MLA meetings,[4] the Library Publishing Forum 2015 meeting,[5] and innovator association meetings, such as VentureWell[6] and DiGRA. e-channel was also featured at the symposium "Teaching and Learning in New Library Spaces: The Changing Landscape of Health Sciences Libraries" sponsored by the National Network of Libraries of Medicine Middle Atlantic Region, the Association of Academic Health Sciences Libraries, and the National Network of Libraries of Medicine Southeastern/Atlantic Region.[7]

Getting more individuals to contribute to e-channel from, it is hoped, around the world is an ongoing goal, as e-channel is intended to be the first place innovators turn to for innovation content regardless of originating institution. Getting the word disseminated about e-channel and deciding how to promote it is key so that a reputation for quality and relevant content is gained. A promotional campaign that reaches innovators and illustrates the value of e-channel to their work is under way. Techniques such as utilizing social media venues, newsletters, websites, meetings and conferences, journal articles, book chapters, and other publication outlets are being employed. Referrals are one of the richest means for spreading the word about e-channel. Encouraging contributors to promote e-channel locally is another means for encouraging use of the platform.

There are many enhancements planned for e-channel besides including more innovators' output. Having an abundance of content will necessitate adding digital object identifiers (DOIs) to the individual pieces in order to enhance discoverability and to provide innovators with a method for referring others to their work, making their content discoverable among the massive quantity of online content. A second enhancement will be to add more metatagging to the content to further enhance discoverability via various search engines. To achieve the goal of providing a platform where innovators can get academic credit for their creations, establishing a peer review process for content is needed.

Sustainability and means of funding e-channel's future are issues that need to be addressed, as, up to this point, everything on e-channel is open access. In time, certain aspects of e-channel will be provided at a fee or potentially via an access fee if development progresses as planned. Funding to date has relied on the EHSL's revenues and budget. Foundation or grant funding will be sought, but a business model for sustainability needs to also be developed. Where the content is stored may also change, as collective cloud space is becoming available that is geared to protecting academic content, such as FigShare. Migration of content will be another sustainability issue as infrastructure software upgrades are made and as new products are released to address new content formats. Preserving the content on e-channel is an added value the EHSL offers innovators, as personal websites (or even institutional ones) can become outdated or content can be removed. Libraries have historically been preservers of knowledge, and multimedia content presents additional challenges for such archiving.

Feedback to date has been positive, and many have expressed that no similar innovation-themed platform exists. Users are pleased with the fresh new look of e-channel that is now WordPress.com based. WordPress.com has enabled the collection of usage statistics so that such metrics can be analyzed to see what content is of most interest to e-channel viewers. Also, usage statistics will provide evidence to funding agencies of the value of e-channel to innovators and others.

CONCLUSION

It is still too early to judge whether the value proposition for e-channel has been achieved. Additional content needs to be added, more promotion achieved, and usage analyzed. Getting university faculty promotion and retention committees to consider e-channel content contributions as viable academic achievements is needed to reach the goal of providing an outlet for innovators for their scholarly outputs. It will be interesting to see if innovators become more comfortable with sharing their intellectual property and innovative process thinking with others as they become more of a component of academic health centers and innovation becomes a true fourth academic mission joining education, research, and clinical care. Will the business culture of academic-based entrepreneurs change over time as they become more embedded in academic settings and work with other academics who are culturally more open with their output? Will innovators and other academic members be willing to give their output to local dissemination venues such as e-channel versus traditional publishers? Will this ease with contributing to nontraditional publishing channels be achieved only once impact metrics are added to such university-based platforms? When will innovators and others who judge their academic merit pivot from considering only traditional publishing venues to include university-based ones? Time will tell, but meanwhile e-channel is forging ahead and is being innovative in its own way to challenge and disrupt academically recognized dissemination venues. Consider contributing to e-channel today by contacting the book's editors.

NOTES

1. Spencer S. Eccles Health Sciences Library, University of Utah, "About Us," 2016, http://library.med.utah.edu/about/?WT.svl=navbarAbout (accessed July 25, 2016).

2. Christy Jarvis, Joan M. Gregory, and Jean P. Shipman, "Books to Bytes at the Speed of Light: A Rapid Health Sciences Collection Transformation," *Collection Management* 39, no. 2–3 (2014): 60–76, doi:10.1080/01462679.2014.910150.

3. Christy Jarvis, Jean P. Shipman, Melissa Rethlefsen, Nancy T. Lombardo, and Tallie Casucci, "e-channel: Forging into the Innovation Dissemination Wilderness" (paper presented at the annual meeting of the Medical Library Association, Austin, TX, May 18, 2015), http://www.mlanet.org/am (accessed May 28, 2016).

4. Jean P. Shipman and Christy Jarvis, "e-channel: Building an Innovation Dissemination Venue" (paper presented at the MCMLA 2015 Virtual Annual Meeting, October 8, 2015), http://www.mcmla.org/2015mtgfiles (accessed May 28, 2016).

5. Christy Jarvis, Nancy T. Lombardo, Jean P. Shipman, and Tallie Casucci, "e-channel: An Innovation Dissemination Venue" (poster presented at Library Publishing Forum 2015, Portland, OR, March 29, 2015).

6. Jean P. Shipman, Tallie Casucci, Christy Jarvis, Nancy Lombardo, Melissa Rethlefsen, and Jeff D. Folsom, "e-channel: An Innovation Dissemination Venue"(poster presented at the VentureWell Open 2015 Conference, Washington, DC, March 20, 2015), http://library.med.utah.edu/e-channel/venturewell-open-conference-2015-posters (accessed May 28, 2016); Barbara Ulmer, Christy Jarvis, Jean P. Shipman, and Andrew Maxwell, "The Innovation Vault:

Advice at Your Fingertips" (poster presented at the VentureWell Open 2016 Conference, Portland, OR, March 4, 2016); Jean P. Shipman, Christy Jarvis, and Chad Johnson, "Calling All Failures" (poster presented at the VentureWell Open 2016 Conference, Portland, OR, March 4, 2016).

7. Jean P. Shipman, "Reducing the Footprint, Expanding the Neighborhood" (panel presentation at NN/LM Joint MAR and SE/A Regions' Teaching and Learning in New Library Spaces: The Changing Landscape of Health Sciences Libraries Conference, Philadelphia, PA, April 18, 2016), http://nnlm.gov/mar/spacesymposium (accessed May 28, 2016).

Chapter Fourteen

Building Innovative Products via Successful Partnerships

Nancy Lombardo and Kathleen Digre

This chapter describes how the Neuro-Ophthalmology Virtual Education Library (NOVEL) (http://novel.utah.edu), a neuro-ophthalmology Web-based discipline-specific repository, was created by the University of Utah (U of U) Spencer S. Eccles Health Sciences Library (EHSL) in partnership with the North American Neuro-Ophthalmology Society (NANOS) to collect, organize, publish, disseminate, and preserve innovative education programming, society literature, and patient education materials worldwide.

NOVEL includes educational publications, journal articles, conference proceedings, a patient education portal, and an *Illustrated Curriculum* of neuro-ophthalmology resources. The unique and innovative collaboration that created NOVEL has evolved and expanded over 15 years (figure 14.1), capitalizing on the strengths of each organization to create valuable scholarly products.

Libraries traditionally collaborate with a wide range of community and professional partners at many levels, usually to increase or enhance access to resources by leveraging membership and funding opportunities. More recently, librarians are stepping into the publishing arena, either providing the platform and technical support for an organization's publications or publishing *gray literature* (i.e., literature produced by organizations but never formally published). In some cases, librarians are creating venues to innovatively publish a wide array of information types (chapter 13). This publishing may include a peer review process, with nontraditional publications being vetted by field experts, which allows authors to receive valid academic credit for their works. This is the case with NOVEL, where the library–society partnership has led to more than simply collections of educational items; it now produces unique educational products of vetted publications in a multitude of formats.

LITERATURE REVIEW

Literature reporting on professional society partnerships with libraries is sparse. Most articles focus on institutional repositories or discipline-specific repositories that collect gray literature and author manuscripts of publications that end up in multiple sources. There was one report from the Texas A&M University Medical Sciences Library and the American Association of Equine Practitioners[1] where information questions as well as conference proceedings of the association are managed by a librarian. A second describes a successful collaboration between a library and a professional society resulting in the development and maintenance of a subject repository of gray literature and conference proceedings in the field of agriculture and applied economics, AgEcon.[2] Wittenberg discusses numerous library publishing ventures that drive innovation and design sustainable models. Teresa Ehling, director of electronic publishing at Cornell University Libraries, explains that librarians are "particularly well positioned to think creatively and knowledgeably about users of digital scholarly material . . . librarians may chart the way in developing new models for delivery of information."[3]

NOVEL'S ORIGIN

NOVEL began as a simple partnership between the EHSL and the Division of Neuro-Ophthalmology (N-O), Department of Ophthalmology, School of Medicine at the U of U. Now, eight committees, from both parties, are involved in producing NOVEL.

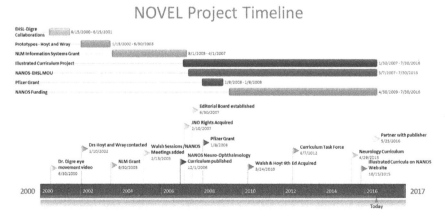

Figure 14.1.   NOVEL Project Timeline.

In 2000, the EHSL established a digital video studio to capture lecture broadcasts. VHS tapes from the Division of N-O were digitized in the studio. These short clips, demonstrating a large number of unique and often rare eye movement disorders, provided new educational content. This material was ideal, as text was inadequate for describing the disorders. The clips were short and broadcast well on the existing technology. The videos were loaded onto a streaming server and made widely available for teaching via a new website.

Kathleen Digre, MD, cocreator of the eye movement collection in the Division of N-O, was in the process of writing a textbook.[4] She wanted to include color slides and videos, which the publisher could not handle. EHSL agreed to create a CD-ROM at no cost to be included with every book, incorporating the color images and videos. EHSL received CD-ROM production credit and the coauthors retained all material rights. This collaborative work solidified the working relationship between the EHSL and the Division of N-O.

Other interesting cases, frequently shared among neuro-ophthalmologists, were identified, and the concept of a subject-specific digital library evolved. Initially, the idea was to focus on slides, videos, animations, and other teaching materials. Dr. Digre, president at the time, approached the NANOS Executive Board to obtain approval for further developing NOVEL with collections from retiring NANOS members. To do this expansion, EHSL offered its technology infrastructure, knowledge management, and project management skills. NANOS members provided the content and the subject expertise for material descriptions. Two prominent members agreed to cull through the work of their careers: one a large collection of 35-mm slides of the optic disc and the other an extensive collection of patient videos representing a large array of neuro-ophthalmic disorders. These two collections formed the prototype for the expanding NOVEL project while grant funding was sought to support and sustain the expansion.

NOVEL FUNDING

Grants were instrumental in launching the NOVEL project by covering start-up costs. The National Library of Medicine (NLM) extramural grant program offered innovative Information Systems Grants at the time to create new and/or unique digital information resources. A joint team submitted a proposal to NLM; the grant was awarded. NLM provided a total of $500,000 over three and a half years, from 2003 to 2007. Additionally, a one-year educational grant of $80,000 was awarded by Pfizer. These grants covered NOVEL costs for the first five years. During this grant-funded period, most of the technology infrastructure and processes for collecting, reviewing, and ingesting content were established. Grant funding covered staffing and technology costs. To launch the project, four staff were funded and hired:

- Principal investigator (librarian): 20 percent—managed the project, oversaw all development, interfaced with the NANOS Executive Board and NANOS committees, and worked with NANOS members on adding new collections, submissions, and reviews
- Project coordinator/graphic designer: 100 percent—worked with EHSL staff to design Web interfaces, manage digitization of materials, and assign metadata
- Video-web technician: 100 percent—digitized and edited videos and assisted with Web interface and design
- Programmer: 10 percent—enhanced the Web interface

Additionally, funds were set aside for consultants to assist with specific parts of the project. These included a metadata cataloger, an audio technician, and a Web developer. EHSL staff were responsible for managing the NOVEL technology. Today, staffing includes a librarian as project manager, with part-time use of technical staff for interface design and part-time clerical staff for managing metadata, uploading, monitoring reviews, and sending out acceptance or revision letters.

Project technology has evolved over time with three major components as well as supplemental tools, including the following:

1. Digital asset management tool—CONTENTdm (CDM), which houses images and files linked to metadata records and provides searching capabilities
2. Video server
3. Website—hosted at EHSL
4. Wiki—the EHSL wiki is used to manage projects, teams, and committees
5. Google tools—help with processes and project management

At the end of the Pfizer grant period, the NANOS Executive Board voted to give EHSL funding annually, with the expectation that personnel and technology would be contributed until additional grant funding was obtained, which was a challenge. EHSL and NANOS entered into a formal agreement to secure NOVEL's future, as it was integral to NANOS' educational goals and to EHSL, as it was a library-produced innovative product. A memorandum of understanding (MOU) was signed on May 7, 2007.

As product development evolved, turning a component of NOVEL into a monetized product was considered. A business plan that described the product, estimated production costs, and identified potential customers was developed, and an agreement with Teton Data Systems, an innovative leader in electronic publishing, was reached to offer this component to libraries, residency programs, and individual subscribers via a licensed product. Generated revenues will be used to further support NOVEL.

PARTNERSHIP AND CONTENT EVOLUTION

An essential component of NOVEL's success was building strong, supportive working relationships with initial content contributors. The contributors devised their own organizational structures and supplied subjects and key words to enhance searching and findability. They provided quality control, making decisions about what to include.

The EHSL NOVEL Team was responsible for converting the contributors' media and knowledge into a usable, navigable digital library. Hundreds of three-quarter-inch tapes were converted to digital video and edited, with direct consultation. Images were scanned from 35-mm slides, some in glass format. The digital files were color corrected. All items were assigned metadata using a standard set of fields. Using proprietary ingestion software, the images and metadata were then uploaded into CDM, the digital asset management system. Videos were mounted separately on a streaming media server and linked to metadata records in CDM. Records also linked to high-resolution TIF image files, stored on an EHSL server. Designed Web interfaces allow browsing and navigation of the collections. All of this work was designed and implemented by librarians and library staff.

Pfizer funding supported the addition of three more collections: (1) a set of classic teaching videos produced professionally by an educational publisher that went out of business, (2) lecture collections, and (3) a collection of materials submitted from all Moran N-O faculty, representing the first collection donated by an institution rather than an individual. Two mid-career NANOS members contributed smaller collections, indicating broader acceptance by the NANOS community.

While the initial collection goals were to develop a set of teaching materials, a number of historical collections became available and were incorporated, including more video, slide collections, and audio lectures. These collections required conversion from old media to digital format, performed by the EHSL NOVEL Team. The video collection was converted from three-quarter-inch or VHS tape and edited with direction from experts. The audio collection was converted from reel-to-reel tape.

In 2005, NOVEL began collecting the presentations from the Walsh Society Sessions, an important meeting that later merged with the NANOS Annual Meetings. The Walsh sessions were the most complex clinical pathologic correlation cases in the discipline. Initially, this collection included meeting abstracts and slide presentations. Later, video captures of the presentations were added. This initial collection of gray literature resulted in NOVEL hosting the entire archive of the Walsh sessions. Similarly, all of the NANOS Syllabi from the complete annual meetings were included. NOVEL began to serve as the archive for conference proceedings and gray literature as well as the educational and historical materials for the Society.

NOVEL continued to procure collections and expand. Rights were secured to the American Academy of Ophthalmology and NANOS Clinical Neuro-Ophthalmology collection produced on CD-ROM in 1999. The 405 images in this collection are of selected cases from the NANOS teaching slide exchange that had been contributed by 30 NANOS members. Metadata were completed for each image and included on the CD-ROM. The images were reviewed by the NANOS Web Education Committee for quality. The review process initiated the development of a NANOS Editorial Board (EB) for peer review of all NOVEL materials.

A patient portal was added as part of the deliverable for the Pfizer grant, adding a set of patient brochures on the more common N-O disorders in multiple languages. Translation into additional languages continues today, with some brochures being in as many as 16 different languages. In addition, each disorder has canned search strategies associated with it to retrieve any NOVEL educational or research materials available on the topic, a link to the topic in *MedlinePlus* and *PubMed*, and links to related support or research groups. The patient portal has been used by physicians and the public.

Another Pfizer grant deliverable created a rare-disease registry within NOVEL. The registry is promoted among NANOS members to encourage practicing N-Os to submit when they encounter new relevant cases. More collections are added over time, including those described in the next sections.

ILLUSTRATED CURRICULUM COLLISION

This addition of collections coincided with a major collaboration between the EHSL NOVEL Team and the NANOS Curriculum Committee that had just completed a final draft of the official NANOS Neuro-Ophthalmology *Curriculum Outline*, which listed every major element required for study by N-O residents and fellows. The EHSL NOVEL Team used this outline to create an overarching organization for all of the NOVEL collections that revealed key missing topical elements. The *Curriculum Outline* was finalized and published in 2006 by Valerie Biousse, MD, chair of the NANOS Curriculum Committee, in the *Journal of Neuro-Ophthalmology* (*JNO*).[5] This iterative collision sparked the creation of an *Illustrated Curriculum*, where every topic in this extensive outline was linked to excellent peer-reviewed examples and teaching materials, such as images, videos, presentations, animations, papers, posters, and working papers.

In 2011, another significant milestone occurred. Until then, the sixth edition of the textbook *Walsh & Hoyt's Clinical Neuro-Ophthalmology* was edited by several NANOS members.[6] In 2011, the publisher returned all rights for the electronic textbook to the editors, two of whom generously offered to give the entire set to NOVEL. This is the primary textbook for the discipline and is an important addition, as it provides background information for much of the *Illustrated Curriculum* as well as an outstanding reference for all NOVEL users.

The NANOS Curriculum Committee decided to improve the scope of the NOVEL collections and fill gaps in the *Illustrated Curriculum* by splitting the entire Walsh and Hoyt textbook into small topics, ingesting these into CDM, and linking these PDFs to appropriate places in the *Curriculum Outline*. The objective was to enhance the *Illustrated Curriculum* by linking textbook sections, premier images and videos, and relevant presentations, papers, or articles from conference proceedings and the *JNO*. The *Illustrated Curriculum* would tie all the disparate parts of NOVEL into a comprehensive educational tool for the discipline.

The second phase of the *Illustrated Curriculum* project assembled a Curriculum Task Force consisting of five NANOS members to review an assigned section of the *Curriculum Outline*. They searched NOVEL to identify appropriate teaching and learning objects for each topic in the outline. Following this intensive effort, existing items were tagged and new items identified or created and added to complete the *Illustrated Curriculum*. After this intensive review, the Task Force members recruited teams of four to five colleagues to continue the effort. After a year, a new assessment of the outline identified remaining gaps. NANOS members are encouraged to submit materials to help fill these identified gaps.

NANOS BOARD AND COMMITTEE CONTRIBUTIONS

The NOVEL EB peer reviews collections and new submissions. It consists of 12 board-certified N-Os—all NANOS members. The EB also did a retrospective review of all collections. The EHSL NOVEL Team designed and developed the electronic procedures and tools used to carry out this review, resulting in some materials' removal. This retrospective review process lasted for about two years.

The NANOS Publications Committee negotiated with the publisher of their journal, the *JNO*, to obtain the rights to archive the journal on the NOVEL site, with a one-year embargo. Lippincott Williams & Wilkins transferred the complete collection of *JNO* electronic archives for the years

1994–2005 in the publisher's PDF format. After further negotiations and a little detective work, the archives of that title were located and transferred to complete the archive of the society's journal. This not only provided an alternative means of accessing and preserving the subject literature but also made access more convenient, as the content was integrated into the search of all NOVEL materials. NOVEL users can now discover teaching materials, gray literature, and the publisher's version of the *JNO* articles when searching NOVEL.

The combination of strong collection building with important NANOS member participation and the active engagement with the NANOS Curriculum and Publication Committees secured NOVEL as an integrated and embedded component of NANOS. NOVEL initiatives were now included in the society's strategic plan. An annual report is presented by the EHSL librarian, Nancy Lombardo, to the NANOS Executive Board at the society's annual meeting. She has become an honorary member of NANOS and serves on the NOVEL Steering Committee, the NOVEL EB, and the NANOS Curriculum Committee.

The Curriculum Committee also tackled establishing a neurology-based N-O curriculum to detail what a neurologist should know about the subject. This Neurology Curriculum was completed in 2015. Plans are in place to add other tiered curricula for students and ophthalmologists in the future. NANOS has added the Illustrated and Neurology Curricula to the member-protected website—a favorite membership benefit.

PEER REVIEW CONTRIBUTIONS

A new set of tools was developed to provide submission instructions, author guidelines, and a methodology for a peer review process for NANOS submissions. Each new submission adheres to author guidelines that indicate the need to provide a standardized set of metadata and descriptions as well as complete author credentials. A new online review form, designed using Google forms, captures the reviews in an online database. With the oversight of the chair of the EB, the EHSL NOVEL Team receives submissions, prepares them for review, compiles a list of items for review, and assigns reviewers small sets of objects to review within two weeks. A major goal of this peer review publishing facet is to offer a highly expedited process to get materials placed within NOVEL quickly. Reviewers can approve, suggest revisions, or reject the objects. If approved, the author receives a formal acceptance letter, signed by the chair of the EB, including a citation for the item, with the NOVEL reference URL. Authors are encouraged to add these citations to their curriculum vitae as proof of peer-reviewed scholarly output and contribution to the knowledge base in their discipline. NOVEL is therefore providing the opportunity to all NANOS members to publish in a peer-reviewed educational platform, which is an attractive membership benefit. Being able to publish also attracts residents and fellows who can demonstrate their initiative and educational contributions during training.

Reviewers are asked to judge if new submissions are of high enough quality to warrant adding them to the *Curriculum Outline*. If affirmative, reviewers indicate where in the outline the submission should be added. All reviewers are aware of the *Illustrated Curriculum* and expedite the filling of gaps by assigning the best submissions to be added to the outline.

NOVEL USAGE

Web statistics are collected for usage of the NOVEL website, though they are not always accurate. CDM has never been able to produce meaningful usage statistics, leaving a huge gap in reporting. Statistics are gathered from the Web interface and video servers, but changes in analytics tools, multiple migrations of video content, and numerous redesigns over the years have resulted in incomplete and inconsistent accounting of actual use (figure 14.2).

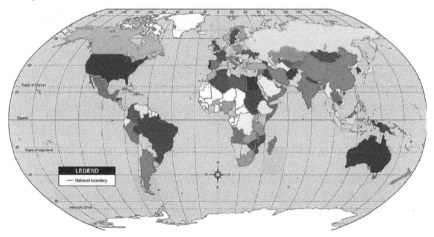

Figure 14.2.   NOVEL Users Originate from Many Countries.

LESSONS LEARNED

The most important lesson learned from this extensive project is that maintaining relationships is the essential key to an innovative collaboration of this scale. This project has endured and persevered because participants are committed and work together to accomplish a shared vision. Establishing an MOU is essential to define roles and key contacts and to outline operational guidelines. Continuous project review via annual reports maintains accountability and informs participants.

A major pivot point in this project was the development and publication of the *Curriculum Outline*. It provided an overarching mechanism for tying the disparate NOVEL collections together. Making this valuable tool available to NANOS members provided a tangible outcome with clear benefits. Providing the opportunity for peer-reviewed publication has generated enthusiasm and increased participation.

Another pivot point was the failure of obtaining a second grant from NLM. Producers were forced to think more broadly, seek alternative funding, and develop a business plan that offered sustainability. This resulted in developing the MOU. A recommendation for any library considering a similar project is to establish an MOU and to consider sustainability from the beginning. It is crucial to identify the primary audience and establish scope and manageable goals. Developing products that serve the needs of members and generate revenue for sustainability is important.

This project demonstrates that libraries can create publishing platforms that give form and function to learning objects, with a resulting product that is systematic and authoritative. Librarians have the expertise to communicate with experts and translate their knowledge into meaningful shared resources. Societies have the content and the subject knowledge necessary to make the objects meaningful. Collaborations between a library and a society can be a win-win, building on the strengths of each and encouraging bidirectional learning. Libraries can gain international recognition and receive income through these partnerships. Using the fundamental expertise of librarians can create new knowledge with original content and expedited publishing.

SUMMARY

NOVEL continues as an open library of N-O resources. By partnering an academic health sciences library with a national society, this project has accomplished far more than either organization could have accomplished alone. Relationships between innovative individuals were essential to the project's start-up, but expanding and nurturing human relationships have led to a network of invested individuals, making the collaboration strong. As more content was added, the level of trust and confidence in the partnership grew. There was an increased willingness by both parties to tackle new tangents. The EHSL NOVEL Team counts on the commitment of NANOS members who volunteer to spearhead new initiatives, and NANOS believes that new projects will be completed by the EHSL. Library–society partnerships are ideal, as they can bring together the complementary expertise of two organizations.

NOVEL has become an integral part of NANOS and is presented at every NANOS annual meeting with updates to the Executive Board and the full membership and at meetings of committees. An information and user assistance table is a common feature. Members discuss current work in progress and how to use NOVEL effectively and learn what is new. They offer suggestions for new areas of growth and are encouraged to contribute submissions. Anecdotes about NOVEL are shared as well. One physician described an event where his patient appeared in his office with numerous NOVEL cases printed from the website. Another neuro-ophthalmologist explained how she had completely changed the way she teaches now that she has NOVEL to support her. Members appreciate this resource and societal benefit.

NOTES

1. Robin R. Sewell, Norma F. Funkhouser, and Christine L. Foster, "Reaching beyond Our Walls: Library Outreach to Veterinary Practitioners," *Journal of Veterinary Medical Education* 38, no. 1 (2011): 16–20, https://www.researchgate.net/publication/51537175_Reaching_Beyond_Our_Walls_Library_Outreach_to_Veterinary_Practitioners.

2. Julia Kelly and Louise Letnes, "Managing the Grey Literature of a Discipline through Collaboration: AgEcon Search," *Resource Sharing and Information Networks* 18, no. 1 (2005): 157–66, https://www.learntechlib.org/p/74041.

3. Kate Wittenberg, "Librarians as Publishers: A New Role in Scholarly Communication," *SEARCHER: The Magazine for Database Professionals* 12, no. 10 (2004): 50–51, http://www.infotoday.com/searcher/nov04/index.shtml (need subscription).

4. Kathleen B. Digre and James J. Corbett, *Practical Viewing of the Optic Disc* (Waltham, MA: Butterworth-Heinemann, 2003).

5. Valerie Biousse, "The Neuro-Ophthalmology Curriculum of the North American Neuro-Ophthalmology Society (NANOS)," *Journal of Neuro-Ophthalmology* 26, no. 4 (2006): 303–15, http://journals.lww.com/jneuro-ophthalmology/Citation/2006/12000/The_Neuro_Ophthalmology_Curriculum_of_the_North.24.aspx.

6. Neil R. Miller, Nancy J. Newman, Valerie Biousse, and John B. Kerrison, eds., *Walsh & Hoyt's Clinical Neuro-Ophthalmology*, 6th ed. (Philadelphia: Lippincott Williams & Wilkins, 2005).

Chapter Fifteen

Educating Innovators

The Innovation Vault

Barbara A. Ulmer and Christy Jarvis

The past decade has seen tremendous growth in the field of innovation and entrepreneurship education at universities throughout the world. Researchers in academia have offered evidence of this phenomenon by documenting increasing numbers of academic programs and faculty positions related to innovation and entrepreneurship since the start of the twenty-first century.[1] In 2006, the Ewing Marion Kauffman Foundation, a nonprofit organization dedicated to advancing entrepreneurship and education, reported a remarkable growth in entrepreneurship education as seen in increased student enrollment, intercollegiate business design competitions, and newly implemented entrepreneurship curricula and programs. The report draws attention to the growing importance of innovation and entrepreneurship in higher education, and it notes that "in the past three decades, formal programs (majors, minors, and certificates) in entrepreneurship have more than quadrupled, from 104 in 1975 to more than 500 in 2006."[2]

The University of Utah (U of U) has joined numerous peer institutions in devoting resources to supporting the evolving needs of innovation and entrepreneurship education. In 2012, the university president declared the institution's commitment to "cutting-edge teaching and research that fosters inter- and trans-disciplinary innovation, creativity, entrepreneurship, and knowledge/technology transfer"[3] and made this central to the university's mission. In practice, this pledge to advance innovation and entrepreneurship has led to the creation of an Entrepreneurial Faculty Scholars program, a collaborative Center for Medical Innovation, and the construction of the Lassonde Studios, where 400 student innovators are empowered to *Live. Create. Launch.*

The explosive growth of academic programs aiming to teach students the unique skills required to be successful innovators and entrepreneurs is accompanied by a growing need for educational content in these areas to be used in curricula. A key component of training students to develop an innovative mind-set is to utilize innovative models in teaching.[4] The *best* methods for teaching students the skills and processes that lead to innovation success are still under debate, but many academic institutions are moving away from deductive classrooms in which the instructor introduces and explains concepts to inductive ones where the instructor shares examples that illustrate the concept and allows students to reflect on how the concept works.[5] Some institutions have experimented with a massive open online course (MOOC) format for teaching entrepreneurship,[6] while others have shifted to a *practice-based* learning model that focuses on learning by doing and stresses practical competencies.[7]

Whatever teaching methodology is employed, however, it is beneficial for students to be exposed to a multitude of attitudes and perspectives on the innovative process. Through interactions with instructors, classmates, and industry leaders—all of whom provide unique points of view and whose collective wisdom is often greater than the sum of its individual components—this can be achieved. To further broaden the field of content experts from whom students can learn, the Spencer S. Eccles Health Sciences Library (EHSL) at the U of U partnered with the Canadian Innovation Centre (CIC) to create the *Innovation Vault*—an organized collection of videos from a wide variety of industries centered on the topic of innovation. From product innovation in aerospace to process innovation in business and social innovations hoping to change the world, the *Innovation Vault* contains lectures, formal classes, orations, and interviews from experts in both the private sector and academia. Topics range from taking a new invention to market and obtaining investors for an entrepreneurial start-up to utilizing the stage gate growth model within existing organizations and cutting-edge research on predicting successes in business innovation. This rich educational video library enables instructors to introduce their students to thought leaders and cutting-edge pioneers in the field of innovation and empowers students to take charge of their own learning processes.

This chapter describes the collaborative effort between the EHSL and CIC that led to the creation of the *Innovation Vault* and discusses EHSL's process of obtaining and organizing the content into a searchable knowledge repository.

BACKGROUND

The EHSL at the U of U is one of three university libraries and serves the Colleges of Health, Nursing, and Pharmacy and the Schools of Dentistry and Medicine as well as the U of U Health Care Hospital and Clinics. It also is the library for many interprofessional centers and institutes. EHSL shares physical space and works collaboratively with several U of U innovation programs (chapter 3) and has become increasingly involved in partnering with campus innovators to preserve and disseminate their scholarly output via EHSL's e-channel platform (chapter 13). EHSL's close working relationship with various U of U innovation programs has resulted in opportunities to engage with innovators from many disciplines by participating in online communities and networking with like-minded peers at innovation-themed conferences. As ideas are shared and solutions to problems are debated, organic partnerships arise between interested parties. One such partnership emerged from a conversation held between the EHSL director and Andrew Maxwell, PhD, of the CIC at the VentureWell Open 2014 Conference in San Jose, California.

The CIC is a national, not-for-profit organization dedicated to helping innovators, inventors, and entrepreneurs transform their ideas into market successes. The group works across multiple industries in the private sector, but it is also deeply involved with academia and government. It works with leading Canadian universities in the areas of innovation education and the commercialization of research. The CIC also works with the government by giving guidance and insight to Canadian policymakers interested in driving innovation forward.[8]

The conversation between EHSL and CIC revealed that Maxwell, the chief information officer and director of partner relations at CIC, had been involved since 2012 in an ongoing project intent on capturing video interviews with luminaries in the field of innovation and entrepreneurship with the goal of creating a resource of current industry and academic knowledge for students at the CIC. As the project grew, he also began capturing video documentation of lectures, seminars, and conference breakout sessions on topics relevant to the study of entrepreneurship and innovation. Once recorded, these videos were edited and then uploaded to the CIC's Vimeo video channel. The challenge of coordinating the hundreds of collected videos into a searchable database offering ease of access and discoverability by content topics for students and faculty was under discussion among Maxwell and his colleagues when Jean Shipman, EHSL director, introduced him to e-channel and proposed a collaboration that would meet this need for organizational structure of the video collection.

METHODS/PROJECT TIME LINE

EHSL faculty and staff had been experimenting with various iterations of the e-channel platform and testing their viability in terms of content, capacity, discovery, and uses for students and faculty. At the point where the CIC videos were made available for download by EHSL, e-channel was in its third iteration (see chapter 13 for specifics on the growth of the platform). This provided EHSL with space to add and initially organize more than 200 of the videos in the CIC collection. To accomplish this, the videos were downloaded from CIC's Vimeo site and transferred to an EHSL-controlled YouTube channel—an *Innovation Archives* playlist that was hidden from public view and used as a holding area for viewing and organizing work.

Once all videos were downloaded to the *Innovation Archives* playlist, an EHSL research associate viewed each video, determined its suitability for inclusion in the library, and assigned those selected to one of five unique categories. New playlists were created within the publicly viewable *Innovation Vault* to correspond to each of these categories as a way to organize and associate similarly themed videos. These five initial playlists were the following:

- Master Class Playlist, which contained substantive interviews of innovation authors discussing their recent works, such as Geoffrey Moore, Michael Raynor, Alan Hevner, and Jeffrey Dyer; professors from various universities involved in innovation and entrepreneurship programs, such as Noam Wasserman, Dean Shepherd, Jacob Goldenberg, and Richard Boyatzis; and serial entrepreneurs and venture capitalists responsible for start-ups in various technology, aviation, and biomedical fields
- Multi-Video Playlist of a single topic series, such as Bob Cooper's 13-video Stage Gate series
- Special Events Playlist of videos including multiple Innova-Con events sponsored by the International Association of Innovation Professionals, multiple conferences and meetings sponsored by the International Society for Professional Innovation Management, events sponsored by the Innovators Alliance network of corporate chief executive officers who foster profitable company growth through innovation, and *Backbone* magazine's *Start Me Up* event from 2013
- Spotlight On Playlist, featuring single video classes and lectures on specific innovation and entrepreneurial topics, such as the Value Proposition Canvas, when to start a business, eight critical factors for venture success, alternate business models for new innovators, and funding source options for new start-ups
- Insider Perspective Playlist, single-video orations on specific innovation or entrepreneurial subjects from members of corporate, academic, and

governmental innovation teams working within an organizational innovation project or department

Once divided into their respective playlists, each video was assigned an individual tag number that indicated its subset playlist and its order for viewing in cases where one was required. Each video was then viewed repeatedly in order to create a succinct title and descriptive overview of the content. These descriptions contained all possible relevant identification information that could be gleaned from the video, including the name and credentials of the speaker, type of video (classroom lecture, oration, or formal interview), interviewer name and credentials where appropriate, date, location, institution associated with the event, and highlights of the content. Also included in this description area were applicable key words taken from a list of more than 48 terms gleaned from the original viewing of the videos, with later additions made as needed.

Initial testing with students and faculty at U of U was done at this point, and the project was met with great enthusiasm. The videos are searchable by any word used in either the title or the description area. For example, users are able to find multiple videos by one person on a recent book release that was discussed, both in a direct interview and again at various conferences, or on a single topic, such as *disruption theory* or *design theory* commented on by multiple experts in various settings.

At this point, the e-channel development team had begun addressing two different platform limitations that had emerged as the site evolved. The first of these concerned a restriction on the number of collections that could be represented on the home page. The second issue was the need to increase searchability of specific topics, key words, speakers, organizations, and events across multiple file formats, such as videos, conference posters, text documents, audio files, and others. These limitations had a direct impact on the usefulness of the *Innovation Vault* content. While efforts were under way to increase the site's display capabilities and improve its search function, the *Innovation Vault* project leader decided that it would be beneficial to group the videos by content topic rather than by presentation format in order to facilitate specific searches as well as enable students and faculty to browse for educational content within a specific innovation topic. This would also achieve maximum exposure for videos covering multiple topics, as these comprehensive videos could be assigned to multiple subject playlists if appropriate. With this goal in mind, the *Innovation Vault* project leader reconfigured the existing video subcategories, created new topical playlists, and reassigned the hundreds of videos to one or more of the newly established subject categories.

OUTCOMES/RESULTS

At the end of 2015, the e-channel development team migrated e-channel content, including the *Innovation Vault* playlists, to a new WordPress.com platform that offered additional content capabilities and enhanced organizational layout and delivery formatting. The new platform offers the ability to group the CIC videos into topical playlists based on key terms that represent important concepts and stages in the innovation life cycle. See table 15.1 for examples from mid-2016 of the various areas of innovation and entrepreneurship covered by these topical playlists.

The topical playlist subset groupings is one of the most popular features of the *Innovation Vault*, as evidenced by the platform's usage statistics. It encourages exploration by students interested in various aspects of innovation. It also allows faculty instructors to augment their curriculum by creating customized collections of recommended or required videos that illustrate various concepts and provide perspectives from innovation luminaries. This accomplishes the goal of exposing students to a multitude of attitudes, opinions, and strategies that can be used for steering the innovation process toward a successful outcome.

Throughout the initial organizational phases of the project, the CIC has continued to gather additional videos, so the process of downloading new content and organizing it within the various topic sets and subsets has continued. As of this writing, there are approximately 350 videos in the e-channel CIC collection. The topical key word list also continues to grow with new subsets being created as appropriate; an ontology of these terms was generated.

CONCLUSION

As innovation and entrepreneurship education programs continue to proliferate in academic institutions around the world, the need for educational material to support learning objectives will only increase. The *Innovation Vault* offers a wealth of knowledge and insider expertise from dozens of luminaries in the field—all organized, described, and disseminated in a way that enables instructors to utilize the content in their classrooms. The online format allows for embedding of videos into existing learning management systems, such as Canvas and Blackboard, and also empowers instructors to explore innovative ways of delivering content, such as possible integration with assigned e-books or the development of self-directed learning modules. Students can also view the videos as they desire on their own accord.

The *Innovation Vault* video library will continue to grow as the field of innovation and entrepreneurship education evolves into an established disci-

Table 15.1. Topical Playlists.

Topical Playlists	Number of Videos
Business Model Canvas	2
Business Model Innovation	12
Commercialization	13
Design	8
Disruption	16
Ethics	25
Founders	32
Funding	14
Gaming	4
Human Elements	92
Ideation	23
Insider Perspectives	30
Intellectual Property	6
Market Systems	23
Master Class	115
Medical Innovations	9
Open Innovation	20
Patents	5
Predicting Success	44
Process Innovation	6
Product Development	24
Sample Pitches	16
Serial Entrepreneurship	14
Social Entrepreneurship	20
Spotlight On	86
Stage Gate	18
Start-Ups	39
Sustainability	15
Tech Innovations	20
Value Proposition	9

pline. The *Innovation Vault*'s potential to become an indispensable knowledge repository for educators, students, and innovators is accomplished through the inclusion of a multitude of perspectives, opinions, and experiences. To make the *Innovation Vault* as robust as possible, consider contributing your own voice to the collection by contacting this book's editors.

NOTES

1. George T. Solomon, Susan Duffy, and Ayman Tarabishy, "The State of Entrepreneurship Education in the United States: A Nationwide Survey and Analysis," *International Journal of Entrepreneurship Education* 1, no. 1 (2002): 65–86, http://www.senatehall.com/entrepreneurship?article=19 (accessed October 26, 2016); Jerome A. Katz, "The Chronology and Intellectual Trajectory of American Entrepreneurship Education 1876–1999," *Journal of Business Venturing* 18, no. 2 (2003): 283–300, doi:10.1016/s0883-9026(02)00098-8 (accessed May 27, 2016); Scott R. Safranski, "The Growth and Advancement of Entrepreneurship in Higher Education: An Environmental Scan," *Academy of Management Learning and Education* 3, no. 3 (2004): 340–42, doi:10.5465/amle.2004.14242270 (accessed May 24, 2016); Donald F. Kuratko, "The Emergence of Entrepreneurship Education: Development, Trends, and Challenges," *Entrepreneurship Theory and Practice* 29, no. 5 (2005): 577–98, doi:10.1111/j.1540-6520.2005.00099.x (accessed May 27, 2016); Matthew M. Mars and Amy Scott Metcalfe, *The Entrepreneurial Domains of American Higher Education* (San Francisco: Wiley/Jossey-Bass, 2009).

2. Ewing Marion Kauffman Foundation, "Entrepreneurship in American Higher Education," 2008, http://www.kauffman.org/what-we-do/research/2013/08/entrepreneurship-in-american-higher-education (accessed May 24, 2016).

3. Office of the President, University of Utah, "7 Core Commitments," 2016, http://president.utah.edu/7-core-commitments (accessed May 24, 2016).

4. Dean A. Shepherd, "Educating Entrepreneurship Students about Emotion and Learning from Failure," *Academy of Management Learning and Education* 3, no. 3 (2004): 274–87, doi:10.5465/amle.2004.14242217 (accessed May 24, 2016).

5. Ikhlaq Sidhu, Ken Singer, Charlotta Johnsson, and Mari Suoranta, "Introducing the Berkeley Method of Entrepreneurship—A Game-Based Teaching Approach," *2015 American Society of Engineering Education Annual Conference and Exposition Proceedings*, 2015, doi:10.18260/p.24367 (accessed June 27, 2016).

6. Mushtak Al-Atabi and Jennifer DeBoer, "Teaching Entrepreneurship Using Massive Open Online Course (MOOC)," *Technovation* 34, no. 4 (2014): 261–64, doi:10.1016/j.technovation.2014.01.006 (accessed June 27, 2016).

7. Tero Montonen and Päivi Eriksson, "Teaching and Learning Innovation Practice: A Case Study from Finland," *International Journal of Human Resources Development and Management* 13, no. 2/3 (2013): 107, doi:10.1504/ijhrdm.2013.055412 (accessed June 27, 2016).

8. Canadian Innovation Centre, "Canadian Innovation Centre," 2016, http://innovationcentre.ca (accessed June 27, 2016).

Chapter Sixteen

Information and Innovation

What Does the Future Hold?

Jean P. Shipman and Barbara A. Ulmer

Predicting the future is never simple, though many fields lend themselves more easily than innovation toward a linear trend progression of current growth patterns and where they could lead in the future. It becomes almost impossible to realistically envision how new approaches and inventions could pivot the innovation field and take the current landscape off into an entirely new direction. The very nature of disruptive innovation precludes the ability to predict the pivot prior to it happening. What can be known, however, is that success in a future climate of increased innovation will require the current silo approaches to innovation in educational institutions and corporate think tanks to give way to combining talents and knowledge previously sectioned off from one another. Libraries are in a pivotal position to help facilitate the need to combine varying talents from different institutional entity types and their corresponding knowledge bases.

With this premise in mind, what will the future of innovation within academic health centers be like, and, concurrently, how will librarians partner with innovators of the future to meet their information needs and to survey the environment in which all will work? That is the goal of this chapter: to offer some likely hypotheses and insights into how these two terms, *information* and *innovation*, will continue to be a natural combination in the years ahead. Up to this point in the book, the authors have provided case studies or examples of how intertwined information and innovation are together. The editors predict that this combination will continue to thrive, as information will be even more valuable to innovators, and here is why.

INNOVATION FUTURE

The recent growth in the trend of innovation has its roots in the early days of the industrial revolution, where innovators such as Ford and Carnegie had impacts on the world that continue to shape lives today. With the onset of the technology boom and the integration of the Internet into mainstream global culture, life continues to morph into previously unrecognized iterations at an ever-increasing rate. The reality of a lifestyle where change increases geometrically is already keenly felt even while it is recognized that the world is still in the infancy of this revolution. Innovations in products, services, processes, and business models are leading to new methods for designing, creating, delivering, discovering, and consuming goods and services. Although this book focuses on technological innovations, particularly in the health and medical arenas, the scope of the impact of this age of innovation has placed the tools for ever-increasing methods and discoveries in the hands of a generation weaned on pixels and megabytes, giving them the vision of a future that lays beyond the imagination of many of those having grown up in a time when disruptive innovation occurred occasionally.

As this innovation growth curve continues to skyrocket, the overall knowledge base of the human race will expand exponentially, making the task of knowledge tracking, organizing, and discoverability more challenging than ever. To remain vital, those in the forefront of innovation will have a greater need to know what their predecessors encountered along the road to success, where detours occurred, and how pivots were handled. This will require novel methods of capturing, storing, and organizing this huge new repository of information. Compared to the concept of librarian as it has been understood for the past 50 years, this will necessitate the transformation of librarians' skill sets to meet the challenges ahead. Librarians will need to practice in concert with innovators of the future and work within their contexts as informationists, providing timely updates on innovation attempts and progresses while cataloging processes as they occur and managing information to successfully move new innovations from concept to market.

Institutional Dependency on Innovation

As clinical revenues and research dollars have become less available through health care reform and reduced government extramural funds, universities, especially academic health centers, have turned to innovation as their new source of expendable income. Innovation leads to commercialization that leads to revenue. Many universities teach entrepreneurship and innovation, and several have sponsored student competitions to encourage such mindsets in students (chapter 10). The growth of VentureWell is one indicator of how many universities of all kinds are developing entrepreneurial curricula for their students and enlisting faculty to help spark interest in innovation. As innovation rescues the financial status of health sciences universities, the other missions, (e.g., education, research, and patient care) will need to incorporate innovation within their processes and goals as well. Innovation will be more than technology based but will permeate how students learn, how patient care is delivered, and how research is conducted.

Universities will need to offer innovative courses and entrepreneurial curricula in order to maintain their reputations and to recruit ambitious and talented students and faculty. Students at a young age participate in hackathons and competitions and expect to continue their training in such areas when they enter college. Universities will need to have the necessary resources and equipment to foster innovation and discoveries. They will need to ensure that students are attracted to what they offer and that they are able to meet innovation accreditation standards of the future.

Global Influences

Innovators are already working together from around the globe. The Games4Health Challenge at the University of Utah had many interprofessional and international teams (chapter 9). These innovative teams work on problems that need solutions. Team value propositions address the benefits these solutions offer. These diverse individuals will be able to tackle wicked world issues, such as utility shortages, violence, global warming, and other major problems. The creative nature of innovators sparks new approaches to ancient problems and it is hoped, new and more efficient solutions. Testing of created medical devices often occurs in countries other than the United States due to fewer restrictions and regulations. The U.S. Food and Drug Administration is slowing down progress,[1] as it is not funded well, and as innovation skyrockets, the demand for approvals increases, whereas the infrastructure support for approvals has not experienced comparable increases in funding. Other countries can learn from the United States and emulate technologies at a faster rate due to kinder government control. Changes in regulatory processes are needed, and time will tell if speedier processing will emerge.

INFORMATION FUTURE

Librarians' Roles with Innovation

This book has illustrated the many ways librarians and information specialists contribute to innovation. Not only do they partner with innovators, but they are innovators themselves as well. They have incorporated technologies as they evolve, and they have tremendous resilience through repurposing skill sets to remain relevant and vital. As digital collections replaced print ones, librarians had to learn not only computer skills but also how to negotiate licenses for electronic resources and how to convert special print collections into digital collections. In addition, they are becoming knowledge managers and are collecting information and knowledge being generated by their institutions and disseminating such through multimedia and Web-based platforms. Tools are being created that librarians are either generating or assisting with their development, such as apps and games. Librarians have also been creating research citation and impact tools.

Traditionally, librarians have been responsible gatekeepers of information, selecting and acquiring quality content, cataloging it to make it easily discoverable, and also storing it on shelves and in digital formats when appropriate. These same processes are being applied with digital content, but it takes a new skill set to achieve the same results. Often, the content needs reformatting into the proper file type, which usually means scanning (optical character recognition) or word processing. Now numerous formats exist, and librarians learn all types of software to accommodate the variety of file formats (e.g., LaTeX). Librarians apply their organizational skills not only to organizing resources of information but also with organizing people and committee outputs. They know what content to preserve, the most efficient means for doing so, and how to disseminate it to reach appropriate audiences with corresponding capabilities. They are still the gatekeepers of information but rapidly are becoming the gatekeepers of knowledge and, potentially, wisdom. Their expertise can be applied to innovation, which has not had an established mechanism for tracking output and products in the same way that research results have been captured and recorded in journal articles and books. Most innovation documentation is ephemeral and not supplied in traditional scholarly communication containers, such as books and journals. Output containers include gaming wrap kits, medical devices, games, apps, prototypes, design sheets, and so on. How these are and should be gathered, described, stored, and retrieved are expert domain areas of librarians.

Innovators' intellectual processes are somewhat unique; learning can occur by following design processes for new products, devices, and apps. Design often includes a team of interprofessional individuals who contribute special expertise, such as engineering, fine arts, graphic arts, business management, and health sciences knowledge. These teams often produce at the speed of light and frequently shift directions based on user feedback, design flaws, and/or regulatory requirements. Not properly documenting these paths, including who was involved in the design and development processes, inhibits replication and recognition. Librarians can become members of these teams, as indicated in chapters 9 through 11, to help guide the recording of effort, the need for such, and the decisions made along the way that create pivot points or alternative pathways. Such documentation is key to avoiding duplication of effort, building on prior knowledge, knowing when to readdress a failure, and giving appropriate attributions for intellectual contributions. Librarians have a special understanding of the attributes of such documentation that will be of value to innovators as they apply for Food and Drug Administration approval and address other regulatory requirements. Librarians can also determine how best to record innovators' contributions to be able to measure their impact and their influential reach. Tracking team members as they progress from being students to being commercial entrepreneurs or designers is something librarians can and should do. They know methods for recording individual progress and for documenting such for academic recognition and credit when it comes to applying for retention, promotion, or tenure. Innovators should consider librarians their personal shoppers and experts at helping to promote individual value and worth. Impact tools are part of a librarians' tool kit, and knowing how to ascertain one's contributions is invaluable. Many librarians are also faculty and can encourage academic faculty affairs units to recognize nontraditional publishing outlets as worthy of academic merit. They know discipline-specific outlets for knowledge and can make recommendations to innovators of where best to place their results for highest visibility. They can indicate the relative value of such outlets to promotion committees.

Innovation Work Space and Libraries of the Future

This book has provided examples of various work spaces created to foster innovation and for innovators to be most effective, many of which reside within health sciences libraries. As innovation becomes more a part of the fabric of institutions and universities, replicative work spaces will be integrated in more campuses. More dormitories, like the Lassonde Studio at the University of Utah, will be constructed for students to be immersed in innovative environments that foster collisions among different types of students with a variety of academic domains and expertise. Information sources will need to be incorporated into these innovative spaces so that information can be easily applied. In fact, libraries and innovation centers can coexist and be administered primarily by library faculty and staff. Librarians are familiar with operating public shared spaces and can transfer this knowledge to shared discovery centers. This coexistence utilizes the expertise of librarians and enables innovators to proceed with innovating. Budgeting, personnel management, event planning, space scheduling, and so on are all attributes librarians can contribute to operating a shared information and innovation facility. This reduces the costs for both entities, as personnel can be shared, thus leading to more institutional efficiencies and savings.

Workplaces will also experience this kind of modality in that teams of employees may be hired to work on a specific problem or task and then be dispersed once that effort is completed. These teams may be global and virtual and will impact the kind of work space needed to be successful. Knowing appropriate tools and group processes for effective virtual group work will be a needed skill set of innovators and librarians of the future.

For teams that do physically meet and work, future work spaces will need to accommodate physical comfort needs as well as be able to connect virtual team members in the same space. These spaces will need recording mechanisms to capture intellectual processes, decisions, and outcomes—a true *living room*. Team members may live in their work spaces for a while and then relocate to the next project pod. Flexible furniture that is mobile and accommodating to different body types and styles of work is needed. Offices will become more living spaces, and Herman Miller's Living Office line of furniture[2] is but one example of the furniture that will be needed by innovators. Many studies are being done to examine individual needs; workplaces are going to have to accommodate work–life balances, as Google[3] and Amazon[4] are currently offering.

Like the innovation workplaces of the future, the tools that innovators use will also become more ubiquitous and embedded. Nanotechnology will enable the Internet of Things to be a reality; clothing and jewelry will have communication and recording devices embedded within them. Recording of conversations will occur in real time to enable easy recall of decisions reached and work assignments. Innovators will experience enhanced mobility, as being tethered to workstations and devices will be passé; they will be able to move freely about. Distance interactions will be common with new tools again facilitating conversations, work flows, and deliberations. Conversations among individuals will be automatically and easily documented by new tools. Groups will be able to converse with one another across the globe and will easily feel like they are sitting right next to each other as holograms and virtual reality become more common in work environments. More innovators will work from home or small offices and not need to be located at universities or within white-tower institutions.

Classes for innovators will be globally delivered and not just within classrooms. Classes may be very modular in that students specializing in innovation may pick and choose to compile their own desired curriculum from a variety of sources. They may be self-taught and interact with educators and others only when needed.

Simultaneously, workplaces will become showcases, and areas designed to advance innovation will have large displays of generated products, high-definition data visualizations, and light and laser exhibits demonstrating the latest gadgets, products, and discoveries. Workplaces will become true discovery and dissemination centers for all to observe as they stroll by.

Libraries will need to be nimble workplaces as well and will need to be able to supply information in a ubiquitous 24/7 manner. Embedding informationists within innovative contexts will offer innovators ready access to expertise and to librarians' subject knowledge. Health sciences librarians no longer have to be located within libraries, as much of the information content is digital and available from any location. Wireless access to data from any point—and eventually wireless electrical power—will enable information to be obtained at any time from multiple kinds of devices, perhaps even embedded chips within one's brain.

SUMMARY: INFORMATION AND INNOVATION—TOGETHER FOREVER

Information shapes innovation, and innovation shapes information; thus, a natural alliance will still preside in the future between these two concepts. Innovators will need to filter through even more information to locate prior art, past efforts, and workable component identification and also to locate relevant mentors, industry partners, and funders. As the volume of information explodes, the ability to navigate through the wealth of information will mean either that innovators will need to learn information discovery and searching skills or, better yet, that they include on their development teams librarians who bring these skill sets to the table. Innovation informationists are the future, as they are more cost effective and place the level of expertise needed at the appropriate level, much like the medical home workforce model. It takes teams to innovate, especially interprofessional teams, and librarians are natural coordinators of teams with their neutral and bird's-eye view of institutional and university activities. They also are cost effective and can save time for others. They have historically contributed to team development and achievements with their expertise for locating the right information at the right time in the needed manner, leaving innovators time to do what they do best—innovate.

NOTES

1. Kira Zalan, "Get Out of the Way of Medical Progress: Public Policy Needs to Adapt to the Growing Speed of Medical Advances," *U.S. News and World Report News*, 2013, http://www.usnews.com/opinion/articles/2013/12/12/regulations-and-outdated-laws-are-slowing-down-medical-advances (accessed May 29, 2016).
2. Herman Miller, Inc., "Living Office," 2016, http://www.hermanmiller.com/solutions/living-office.html (accessed May 29, 2016).
3. Google, "Our Culture," 2016, https://www.google.com/about/company/facts/culture (accessed May 29, 2016).
4. Amazon, "Our Urban Campus," 2016, https://www.amazon.com/p/feature/4kc8ovgnyf996yn?ref_=aa_tbbx_all_10&pf_rd_r=K13HXQF3M1PN8A9W19M0&pf_rd_p=a506cbab-5ff8-4775-8b46-179ac8523280 (accessed May 29, 2016).

Index

Altziger Jr., Roger, 77–88, 131–139, 189
Arches Saves Your Bacon, 82, 93, 136

B-2-B competition. *See* Bench to Bedside competition
Bench to Bedside competition, 18, 100, 101, 102, 104, 109, 121
BioInnovate, 123
Biomedical literature searching, 107–109
BMC. *See* Business Model Canvas
BMEidea competition, 40
Business Model Canvas, 14, 16, 19, 20, 103, 122

concept development. *See* ideation
Canadian Innovation Centre, 170, 171
Casucci, Tallie, 13–21, 89–97, 99–114, 117–128, 190
Center for Medical Innovation, 19, 23–24, 120
CIC. *See* Canadian Innovation Centre
CMI. *See* Center for Medical Innovation
colleague, 16, 125
commercialization, 13, 19, 20, 23, 80, 95, 101, 126, 171, 174, 178. *See also* marketing
Curriculum Outline, 162–163, 166

design, 3, 18, 25–28, 43, 45–46
Design Box, 15, 134, 136–137
DiGRA, 127

Digre, Kathleen, 157–167, 190
disruption, 1, 173
Doodle Health, 125
Drawing Health, 125

e-channel, 82, 119, 138, 143–154, 172, 173–174
Eccles Health Sciences Library (EHSL). *See* Spencer S. Eccles Health Sciences Library
EHSL director, 24, 25, 26, 29, 30, 56, 90, 93, 96, 117, 118, 128, 131, 133, 171. *See also* information transfer, director of information transfer (DIT)
electronic health records, 133, 134. *See also* Open EHR App
electronic medical records. *See* electronic health records
Entrepreneurial Faculty Scholars, 16

fabrication laboratory, 17; access to, 27, 63, 65; design of, 24–27, 56, 85. *See also* Gary L. Crocker Innovation and Discovery Laboratory; Lassonde Studios
failure: mindful, xiii, 152; mindless, xiii
Faust Law Library, 100
Felix, Elliot, 75, 190
Ferrill, Thomas (TJ), 99–114
Fogg, Benjamin (Ben), 99–114, 190
funding, 18, 21, 93–94, 113, 159–160

Games4Health Challenge, 90, 95–96
gaming, 77–78, 80, 81–82, 87–88, 91, 131. *See also* medical digital therapeutics
Gary L. Crocker Innovation and Discovery Laboratory, 17, 19, 29, 55–65

Herron, Jennifer, 33–41, 191

I-Corps, 18, 121–122
i-Space. *See* Innovation Space
ideation, 13–14, 110–111. *See also Design Box*
Ideation Studio, 15, 111, 139
Indiana University School of Medicine, 33–34
information and innovation, 5, 23, 33, 50, 183; history of, 1–10; space design, 27–28. *See also* librarian roles; library as innovation space; library partnerships
information transfer, 118–128, 138; director of information transfer (DIT), 118–119. *See also* EHSL director; Scholarly communication
innovation, 4, 178; diffusion of, 138; education for, 15, 121–122, 124–125, 169–174, 178–179, 182; information needs of innovators, 82, 91–92; tools and equipment for, 182
Innovation Corps. *See* I-Corps
Innovation Ecosystem, 17
Innovation Guide, 15, 102, 103, 108, 122
innovation librarian, 84, 90–91, 117–118; See also librarian roles
innovation life cycle, 13, 110, 126. *See also* Commercialization; Ideation; Product; Prototyping
innovation process. *See* Innovation life cycle
innovation space, 47–50
The Innovation Vault, 170–174
intellectual property, 23, 77, 100, 106, 112–113, 144, 152, 154, 174. *See also* patent searching
IQ Wall, 36, 38, 40

J. Willard Marriot Library. *See* Marriott Library

James E. Faust Law Library. *See* Faust Law Library
Jarvis, Christy, 143–154, 169–174, 191
Johnson, Chad L., 143–154, 191
Jones, Peter, 99–114, 191

Kaneshiro, Kellie, 33, 191
knowledge commons, 4

Lassonde Studios, 17, 114, 128
librarian roles, 16, 40, 49, 83–84, 86, 87, 89, 90–91, 92–93, 179–180. *See also* EHSL director; information transfer, director of information transfer; innovation librarian; library partnerships
libraries and innovation. *See* information and innovation
Libraries Innovation Team, 100–105
library as collaborative space, 1–2, 4, 9–10, 23, 43–44, 110. *See also* information and innovation; library partnerships; librarian roles; makerspace; synapse
library as innovative space, 103–104, 181, 183. *See also* information and innovation; library partnerships; librarian roles
library as place, 3–4, 6–8
Library partnerships, xiv, 1, 5, 52–53, 74–75, 82, 85, 114, 157, 177. *See also* information and innovation; library as collaborative space; library as innovative space
Lombardo, Nancy, 157–167, 192
Lucia, Joseph, 1–10, 192

maker program. *See* makerspace services of
makerhealth, 40
makerspace, 9, 67–75; access to, 73–74; design of, 36–38, 69; in libraries, 67, 68–69; services of, 38–39; types of, 70. *See also* fabrication laboratory; Ideation Studio; innovation space; makerhealth; Nexus Collaborative Learning Lab; Synapse
marketing, 106–107. *See also* commercialization
Marriott Library, 17, 19, 20–21, 99, 103

Materials ConneXion database, 17
MDT. *See* medical digital therapeutics
medical digital therapeutics, 124–125, 131; patient-centered software, 133–136. *See also* BMEidea competition; DiGRA; gaming; Games4Health Challenge; software development; The Gapp Lab
medical games and apps. *See* medical digital therapeutics
MediGarden, 133–134
mentors, 14, 16
mHealth. *See* medical digital therapeutics
Mirfakhrai, Mohammad, 99–114, 192
Morrison, David (Dave), 99–114, 192
Mowdood Alfred, 99–114, 192

NANOS. *See* North American Neuro-Ophthalmology Society
Neuro-Ophthalmology Virtual Education Library, 157–167
Nexus Collaborative Learning Lab, 33–41
North American Neuro-Ophthalmology Society, 157, 159, 160, 161–162, 162–164
NOVEL. *See* Neuro-Ophthalmology Virtual Education Library

Open EHR App, 92

patent searching, 14, 16, 105–106, 112. *See also* commercialization
patient-facing software. *See* medical digital therapeutics, patient-centered software
Pickett, Timothy (Tim), 55–65, 192
Presentation Practice Studios, 44–47
product, 18–19, 182; regulation of, 19, 113. *See also* commercialization; prototyping; software development
prototyping, 17, 55, 113; additive, 55, 58–59; electrical, 56, 60; information support, 103; rapid, 60; subtractive, 55, 59; tools and equipment, 60–62. *See also* product; software development

Reed, Jacob, 99–114, 192
Rhodes, Nathaniel (Nate), 99–114, 192
Ruth Lilly Medical Library, 33

Schmick, Darell, 99–114, 193

scholarly communication, 143–144, 164–165, 166. *See also* information transfer
Shipman, Jean P., xvii–xxiii, 13–21, 23–30, 55–65, 117–128, 177–183, 189
Skills Center, 27, 29
software development, 139, 140
Spark, 25
Spencer S. Eccles Health Sciences Library, 15, 19, 20–21, 23–30, 56–57, 77, 89–90, 99, 117, 144–145, 157, 171
Stoa, 2
sustainability, 47, 51–52, 64, 153, 166
Synapse, 15, 23–30

Technology and Venture Commercialization, 20, 123–124
Technology innovation life cycle. *See* Innovation life cycle
Technology transfer office. *See* Technology and Venture Commercialization
Tetra-Coalition initiative, 134–135

The GApp Lab, 24, 77, 79–82, 90, 131–133, 139; information needs, 92–95

Therapeutic Games and Apps Laboratory. *See* The GApp Lab
3D printing, 34–35, 47, 50, 58, 59
Tooey, Mary Joan (M. J.), 43–53, 193
TVC. *See* Technology and Venture Commercialization

U-Bar, xiv, 97, 138
Ulmer, Barbara A., xvii–xxiii, 169–174, 177–183, 189
University of Maryland HS/HSL, 43–53
University of Utah: innovation history, 101; libraries, 99–100, 169; support for innovation, 14–16, 17–18, 19–20, 100, 109. *See also* Bench to Bedside competition; Center for Medical Innovation; colleague; e-channel; Entrepreneurial Faculty Scholars; Gary L Crocker Innovation and Discovery Lab; Spencer S. Eccles Health Sciences Library; The GApp Lab

VentureWell, 40, 126

Walker, Spencer W., 13–21, 193
Wasden Christopher (Chris), xiii–xiv, 193

Wimmer, Erin, 99–114, 194
Woodbury, David, 67–75, 194

Zagal, José, 77–88, 131–139, 194

About the Editors and Contributors

Jean P. Shipman is executive director, Knowledge Management and Spencer S. Eccles Health Sciences Library; director of the MidContinental Region and National Training Office of the National Network of Libraries of Medicine; and director for information transfer, Center for Medical Innovation. She is also adjunct professor, Department of Biomedical Informatics, School of Medicine, all at the University of Utah. She is a fellow of and has served as president of the Medical Library Association (MLA) for 2006–2007 and promoted health literacy as her primary presidential initiative. She received the MLA Ida and George Eliot Prize Award in 2006. She also co-convened the Chicago Collaborative and served as a member of the Board of Directors of the Society for Scholarly Publishing from 2013 to 2016. Her professional interests include health literacy, scholarly communications, library administration, innovation, and Lean. She may be contacted at mjshipman@msn.com.

Barbara A. Ulmer has experience as the managing editor of a small publishing company, which, along with her background in financial and technologically innovative process analysis, led to her cofounding and acting as chief financial officer of a technology start-up company in the early days of e-book adoption and her involvement in growing social innovation. She may be contacted at barbara_ulmer@icloud.com.

* * *

Roger Altizer Jr. is the cofounder of Entertainment Arts and Engineering, the top-ranked game design program in the nation; the director of digital medicine for the Center for Medical Innovation; the founding director of The

GApp Lab (Therapeutic Games and Apps); and former director of the Center for Interdisciplinary Art and Technology at the University of Utah. He earned his PhD in communication at Utah and specializes in serious and medical games as well as game design and development. Creator of *The Design Box*, a participatory, inductive design methodology, he works to include users as designers in his work. A former games journalist, he is an internationally recognized speaker who has presented at industry conferences, such as the Games Developer Conference and Penny Arcade Expo, and academic conferences, including the Digital Games Research Association and Foundations of Digital Games.

Tallie Casucci is the innovation librarian at the Spencer S. Eccles Health Sciences Library at the University of Utah. She partners with innovators and entrepreneurs to discover evidence-based knowledge, competitive intelligence, prior art, and potential partners for the creation of emerging technologies, such as novel medical devices, games, and apps. She earned her MLIS at the University of Pittsburgh; a BS in exercise and sport science at the University of North Carolina, Chapel Hill; and a University of North Carolina Business Essentials Certificate from the Kenan-Flagler Business School.

Kathleen Digre, MD, is a professor of neurology and ophthalmology at the University of Utah. She is the past president of the North American Neuro-Ophthalmology Society (NANOS). She helped to establish the Neuro-Ophthalmology Virtual Educational Library (NOVEL) and serves as the chair of the NOVEL Committee for NANOS.

Elliot Felix founded and leads brightspot, a consultancy firm that designs engaging experiences for employees, customers, students, and other audiences and also transforms spaces, services, and organizations. He is a strategist, facilitator, and sense maker who has directed projects for leading companies, universities, and cultural institutions.

Thomas J. (T. J.) Ferrill is the assistant head of creative spaces at the J. Willard Marriott Library. As a part of the Creativity & Innovation Services department, his primary responsibilities are providing support and training for 3D printing, scanning, and modeling for library patrons.

Benjamin Fogg graduated from the University of Utah with a BS in exercise science and an MS in bioengineering and is currently a medical student at the University of Utah School of Medicine. His hands-on interaction with patients, scientific research, and entrepreneurial endeavors are at the core of his passion for medical technology. He is determined to make an impact as he applies his medical experience to the innovation and commercialization of new products.

Jennifer Herron is the emerging technologies librarian for the Ruth Lilly Medical Library, Indiana University School of Medicine. She graduated with her MLIS from Wayne State University in 2012 and completed her degree online, letting her test out a variety of education technology tools while learning the ins and outs of library science. Current projects include promoting mobile resources, 3D printing, engaging library users online via social media, and expanding the library's online presence.

Christy Jarvis has worked in libraries, in various capacities, since 1998. She obtained her MLIS from San Jose State University in 2004. She joined the faculty of the Spencer S. Eccles Health Sciences Library in 2011 as the head of information resources and digital initiatives. In this role, she oversees both traditional and emerging mechanisms for information storage, retrieval, and dissemination.

Chad L. Johnson is the e-channel content specialist for the Spencer S. Eccles Health Sciences Library. He graduated from the University of Utah with a dual emphasis BFA degree in sculpture and photography. Since graduation, he has worked as a photographer's assistant, a photographer, and a visual literacy instructor for the Children's Media Workshop. He spent a decade in Utah's video game industry as an artist, animator, 3D modeler, and art team manager working on many top-tier games. After completing a University of Utah certification program for website design/coding, he was employed by the public radio station KUER FM 90.1 as the producer of its website, http://kuer.org.

Peter Jones, MSLS, spent 15 years in public and academic libraries as a librarian, library manager, and research concierge/informationist. His experience serving as the research concierge for the University of Utah's Center for Clinical and Translational Science and as a consultant for the Center for Medical Innovation led him to an interest in clinical research, statistics, epidemiology, and scientific programming. He currently serves as the statistical data analyst for the Clinical Epidemiology and Infectious Diseases Division of Intermountain Healthcare in Salt Lake City, Utah, and is pursuing an advanced degree in statistics and biomedical informatics.

Kellie Kaneshiro received her BS from Purdue University in microbiology and her AMLS from the University of Michigan. She started her career at the Houston Academy of Medicine–Texas Medical Center Library in Houston before moving to the Ruth Lilly Medical Library at the Indiana University

School of Medicine's Indianapolis campus. Currently, she is the library technology team leader and liaison librarian to several medical school departments and is involved with the undergraduate curriculum renewal process.

Nancy Lombardo has worked in libraries since 1981. She received her MLS from Emporia State University in 1994. She is now the associate director for digital collections. She is the project director of the Neuro-Ophthalmology Virtual Education Library (NOVEL), an educational digital repository that is a collaborative project between the Eccles Health Sciences Library and the North American Neuro-Ophthalmology Society.

Joseph Lucia is dean of libraries at Temple University. Prior to that, he served as university librarian at Villanova University for 11 years. During his tenure at Villanova, Falvey Library won the 2013 ACRL Excellence Award in the "University" category. He served as a member of the Lyrasis Board of Trustees from 2009 to 2012, where he was founding president during 2009–2010, and he currently serves on the board of PALCI.

Mohammad Mirfakhrai holds an EdD with a concentration in human resources management. He is a business specialist at the J. Willard Marriott Library, University of Utah, with 28 years of employment.

David (Dave) Morrison has been the government information librarian and patent and trademark specialist for the University of Utah's J. Willard Marriott Library since January 1987. He is the library's federal depository library representative to the U.S. Government Publishing Office and the Patent and Trademark Resource Center representative to the USPTO.

Alfred Mowdood has been a librarian at the J. Willard Marriott Library at the University of Utah for more than 15 years. He is also an adjunct assistant professor in exercise and sport science.

Timothy (Tim) Pickett, PE, is the director of engineering at the University of Utah Center for Medical Innovation. He is also the cofounder of two medical device companies. He has several patent applications that are pending. His aim is to bring safe, affordable, and effective health care to resource-poor areas across the globe in the areas of women's and pediatric health. As an educator, his biggest reward is to help imaginations come alive through prototyping.

Jacob Reed is a Web application developer and project manager at the J. Willard Marriott Library at the University of Utah and has been creating tools to strengthen and automate library services for the past nine years. He is currently working alongside his colleagues to develop a library-wide digital asset management and preservation system. He is also completing a master's degree in public administration at the University of Utah.

Nathaniel (Nate) Rhodes, MS, is the founder and chief executive officer for Veritas Medical, an early stage medical device company. Nate is an experienced project leader in biotechnology space with 10 years of related R&D engineering and clinical experience. He has won multiple grants and awards from NASA, NSF, and the USPTO. He has published multiple peer reviewed articles, patents, and medical design blogs.

Darell Schmick is the research librarian at the Spencer S. Eccles Health Sciences Library, University of Utah. In this role, he works with researchers, clinicians, and innovators along all aspects of the research life cycle. He received his MLS from Emporia State University and a graduate certificate in organizational change from the Truman School of Public Affairs at the University of Missouri.

Mary Joan (M. J.) Tooey is associate vice president of academic affairs, and executive director of the Health Sciences and Human Services Library at the University of Maryland. She is also the director of the National Network of Libraries of Medicine's Southeastern/Atlantic Regional Medical Library and the National DOCLINE Coordinating Office. She served as president of the Medical Library Association from 2005 to 2006, is a fellow of the association, and was the 2016 Janet Doe Lecturer. She also served as president of the Association of Academic Health Sciences Libraries from 2012 to 2013. She is the 2011 recipient of the Distinguished Alumni Award from the University of Pittsburgh School of Information Sciences. She is the author or coauthor of more than 100 chapters, articles, presentations, or posters. Her professional interests include leadership, emerging trends, library innovation and design, ethics, and mentoring.

Spencer W. Walker, MSc, is the director of regulatory affairs at the University of Utah's Center for Medical Innovation where he teaches and consults for new university-based medical device start-ups. He earned his master's degree in translational medicine from Cranfield University in the UK. He has more than sixteen years of experience in the medical industry as a regulatory affairs and quality executive and consultant/advisor, including Medical Device Regulatory/Quality Compliance.

Christopher Wasden, PhD, is the executive director of the Sorenson Center for Discovery and Innovation at the University of Utah and author of the book *Tension—The Energy of Innovation*, which outlines his approach to innovating technologies and business models. His mission is to enable organizations, leaders, and audiences to regain their creative genius by creating and harnessing the powerful tensions necessary to power the innovation process. He does this by engaging others to ride the innovation cycle along its entire life cycle to discover, incubate, accelerate, and scale their innovative ideas and business models.

Erin Wimmer joined the faculty at the Spencer S. Eccles Health Sciences Library in September 2013 as the teaching and learning librarian, coordinating and administering the library's formal educational efforts. Working with other library faculty and staff, she develops and teaches both curriculum-based and single-session classes. She graduated from the University of Utah with a BA in English in 2007 and completed a dual master's program in library and information science and political science at the University of Southern Mississippi in 2012.

David Woodbury is the associate head of user experience at the North Carolina State University (NCSU) Libraries. He manages technology-rich learning spaces at the libraries, including makerspaces, digital media labs, virtual reality exploration spaces, and collaborative computing areas. He leads several key initiatives, including an expansive student-focused workshop series and the NCSU Libraries' technology lending program. He was a member of the Learning Space Toolkit development team (http://www.learningspacetoolkit.org), an IMLS-funded resource to assist planners with informal learning-space design projects. He received his MSIS from the University of North Carolina, Chapel Hill, School of Information and Library Science.

José Zagal, PhD, is an associate professor with the University of Utah's nationally ranked Entertainment Arts and Engineering program. He has more than 10 years of experience teaching game design and development. He wrote *Ludoliteracy* (2010) and edited *The Videogame Ethics Reader* (2012). His next book (2017) examines role-playing games across different media and disciplinary contexts. He received his PhD in computer science from the Georgia Institute of Technology in 2008.

Lightning Source UK Ltd.
Milton Keynes UK
UKHW01n1949221018
330987UK00001B/5/P